WHAT IS MAN, O LORD?
THE HUMAN PERSON
IN A BIOTECH AGE

Contributors

Peter J. Cataldo, Ph.D.
Director of Research
  and Ethicist
National Catholic Bioethics
  Center
Boston, Massachusetts

Marilyn E. Coors, Ph.D.
Assistant Professor
  of Psychiatry
University of Colorado
  Health Sciences Center
Denver, Colorado

Elizabeth Fox-Genovese,
  Ph.D.
Eleonore Raoul Professor of
  Humanities
Department of History
Emory University
Altanta, Georgia

Mary C. Geach, M.A. (Oxon.),
  Ph.D. (Cantab.)
London, United Kingdom

Luke Gormally, Lic. Phil.
Senior Research Fellow
The Linacre Centre for
  Healthcare Ethics
London, United Kingdom
Research Professor
Ave Maria School of Law
Ann Arbor, Michigan

William B. Hurlbut, M.D.
Consulting Professor
  in Human Biology
Stanford University
Woodside, California

Mark F. Johnson, Ph.D.
Assistant Professor
Department of Theology
Marquette University
Milwaukee, Wisconsin

William E. May, Ph.D.
Michael J. McGivney
  Professor of Moral Theology
John Paul II Institute for
  Studies on Marriage and
  Family at The Catholic
  University of America
Washington, D.C.

Ralph P. Miech, M.D., Ph.D.
Professor Emeritus
  of Molecular Pharmacology
Brown University
  School of Medicine
Division of Biology
  and Medicine
Providence, Rhode Island

Rev. Albert Moraczewski,
  O.P., Ph.D., S.T.M.
President Emeritus
National Catholic Bioethics
  Center
Boston, Massachusetts

John M. Opitz, M.D.
Professor of Pediatrics,
  Human Genetics,
  Obstetrics, and Gynecology
University of Utah
  School of Medicine
Salt Lake City, Utah

Edmund D. Pellegrino, M.D.,
  M.A.C.P.
John Carroll Professor of
  Medicine and Medical
  Ethics
Center for Clinical Bioethics
Georgetown University
  Medical Center
Washington, D.C.

Susan Schmerler, M.S., J.D.,
  C.G.C.
Supervisor
Section of Genetics
St. Joseph's Hospital
  and Medical Center
Paterson, New Jersey

Msgr. Robert Sokolowski,
  Ph.D., S.T.D.
School of Philosophy
The Catholic University of
  America
Washington, D.C.

Rev. William A. Wallace, O.P.,
  Ph.D., S.T.D.
Department of Philosophy
College of Arts
  and Humanities
University of Maryland
College Park, Maryland

WHAT IS MAN, O LORD?
THE HUMAN PERSON IN A BIOTECH AGE

Eighteenth
Workshop for Bishops

Edward J. Furton, M.A., Ph.D.
Editor

Louise A. Mitchell, M.T.S.
Assistant Editor

Cover Design:
Thomas Gannoe

Cover Art:
Michaelangelo
Detail from the Sistine Chapel ceiling

Special thanks to Veronica McLoud Dort
and Amanda Parish

Library of Congress Cataloging-in-Publication Data

Workshop for Bishops of the United States and Canada (18th : 2001 : Dallas, Tex.)
What is man, O Lord? the human person in a biotech age : Eighteenth Workshop for Bishops / Edward J. Furton, editor ; Louise A. Mitchell, assistant editor.
    p. cm.
Includes bibliographical references.
ISBN 0-935372-45-8
1. Genetic engineering--Religious aspects--Catholic Church. 2. Human reproductive technology--Religious aspects--Catholic Church. I. Furton, Edward James. II. Mitchell, Louise A. (Louise Annette) III. Title.

TP248.6.W675 2002
241'.64957--dc21

2002032550

The Human Person

Genetics and Embryology

Disputed Questions

From the Vatican

Angelo Cardinal Sodano

January 20, 2001

Dear Bishop Grahmann,

The Holy Father was pleased to learn that on February 5–9, 2001, the Eighteenth Workshop for Bishops sponsored by The National Catholic Bioethics Center will be held in the Diocese of Dallas. He asks you to convey to all present his cordial greetings and the assurance of his closeness in prayer. He likewise expresses once more his deep gratitude to the Knights of Columbus for their generous support of this important pastoral initiative.

As the Church enters the Third Millennium renewed by her celebration of the Great Jubilee, she senses the need to contemplate ever more fully the face of Jesus Christ, the new Adam, who fully reveals man to himself and discloses humanity's sublime vocation (see *Nova millennia ineunte,* n. 16). For this reason, His Holiness is pleased that the Workshop has taken as its theme: "'What Is Man, 0 Lord?' The Human Person in a

1

Biotech Age." At a time of rapid and at times disturbing developments in the life sciences, bishops are increasingly called to
exercise an ethical discernment grounded in a theological vision of the dignity of the human person, created in the image of
God and destined to fullness of life in Christ.

The advances in biotechnology which have made it possible to decipher the human genetic code, to alter the very biological structures of human life, and even to mix genetic material of different species have opened new vistas for the cure of
human diseases and disorders. At the same time it must be
acknowledged that much of this research raises grave moral
questions, especially with regard to the treatment given to human beings at the first stages of their development. The Catholic tradition, drawing upon the principles of the natural law,
clearly affirms that "the human being is to be respected and
treated as a person from the first moment of conception; and
therefore, from that same moment his rights as a person must
be recognized, among which in the first place is the inviolable
right of every human being to life" (Donum vitae, I, n. 1). Since
the fundamental ethical criterion governing scientific research
can only be the defense and the promotion of the integral good
of the human person, it follows that any procedure performed
on human beings, even at the very dawn of their personal existence, must respect the dignity and rights originating in human nature itself. Far from an extrinsic limitation on human
freedom, this moral obligation arises from the very truth about
the human person.

These principles are particularly relevant in view of the
complex issues raised by the practice of genetic manipulation,
and more so when this gives rise either to discrimination between human subjects on the basis of possible genetic defects
or, more tragically, to the wholesale destruction of laboratory-
produced human embryos. As part of her proclamation of the
Gospel, the Church cannot fail to insist that a child is the supreme gift of marriage, that every child has a right to be born of
a husband and wife, and that every child has the right to be
respected as a person from the moment of conception. Morally
illicit procedures of in vitro fertilization have led to the reprehensible practice of cloning or engendering human beings not
for life, but so that their cells or organs can be obtained and
used for the advantage of others. These are developments which
must be repudiated by any civilized society and above all by the
followers of Christ.

2

In the face of these complex and troubling ethical issues, the Church's Bishops are called to manifest with clarity and conviction an irrevocable commitment to the Gospel of Life. In the end, the discernment of sound ethical criteria and their application to scientific research must be part of a great effort to restore an appreciation of the awesome dignity of the human person, whatever the situation. A great cultural transformation is heeded, in which human beings will be viewed not merely as a complex of tissues, organs, and functions, but as a mystery to be approached with wonder and reverence. This challenging work of education and the formation of consciences with regard to the incomparable worth of every human life demands a sound catechesis capable of communicating the Christian vision of man's creation in the image of God, and emphasizing the value of life and freedom as gifts given for the upbuilding of the individual, society, and all creation in the light of God's plan. Above all, there is a need for constant prayer and the concerted efforts of all members of the Christian community, especially those involved in the fields of science, health care, politics, and ethics, to defend human dignity and to disown the evil of practices and laws hostile to life (see *Evangelium vitae,* n. 100).

The Holy Father is confident that these days of study and reflection will assist the bishops taking part in the workshop in bringing to bear on the profound ethical issues arising in microbiology and the life sciences in our day the beauty and the wisdom of the Catholic tradition. He prays that through the intercession of Mary, Mother of the Church, they will find inspiration and strength for their efforts to proclaim the Gospel of Life and spread the civilization of love. To all present he cordially imparts his Apostolic Blessing as a pledge of joy and peace in the Lord.

With my own best wishes for the success of the Workshop, I remain

Sincerely yours in Christ,

A. Card. Sodano

Secretary of State

GREETINGS FROM THE KNIGHTS OF COLUMBUS

DALLAS, TEXAS
FEBRUARY 5–9, 2001

Your Eminences and Your Excellencies, Reverend Fathers, and Dr. Haas: it is a pleasure to be here this evening to welcome you on behalf of the Knights of Columbus. I have had three occasions to present papers to various workshop meetings in the past: the first while serving in the U.S. Department of Health and Human Services and twice while teaching at the John Paul II Institute for Studies on Marriage and Family.

Tonight I am especially pleased to be here as Supreme Knight of the Knights of Columbus. Since 1980, the Knights of Columbus has provided more than $5.4 million to the National Catholic Bioethics Center to make possible these eighteen workshops for members of the hierarchy. And we have been proud to do so.

One reason that we are proud of this partnership can be found in the Holy Father's address to the Vatican diplomatic corps just several weeks ago. On that occasion the Pope stated, "History will judge [the century just ended] to be the century that saw the greatest conquests of science and technology, but also as the time when human life was despised in the crudest ways." For two decades these workshops have provided scholarly analysis of "the conquests of science and technology" to find ways in which they may be used to advance rather than assault human dignity.

Today this assault continues in a public context that, as we all know, has been conditioned by critics of the Church. The media almost always portray the critics of Church teaching as the ones advancing the cause of human dignity and freedom. Church teaching is portrayed as an archaic list of prohibitions that constitute an irrational obstacle to true human happiness. In other words, the public debate too often transforms the saving Gospel of Life into a moralistic version of "Robert's Rules of Order."

In *Evangelium vitae*, the Holy Father points a way out of this rhetorical thicket. He insists that the "Good News" of the Gospel is the only understanding of the human person and his liberty that is completely consistent with human dignity. And he maintains that man's true calling both as an individual and as part of a community is to live the "Gospel of Life" while building a "Culture of Life." For over twenty years John Paul II has worked to focus the discourse of the "biotech age" precisely on its most profound points: what is the human person and how do we assure both his dignity and his liberty. The topic for this workshop, "What is man, O Lord?" could not be more timely or important.

Permit me one final observation as to why this year's topic is of particular interest to the Knights of Columbus and the mission given to it by its founder, the servant of God Fr. Michael J. McGivney.

Fr. McGivney lived during a time in American history when the social and economic theories of radical individualism, social Darwinism, and survival of the fittest were at a high point. In founding an association of men in such a cultural climate one might have expected those first Knights to seek society's acceptance by championing those very ideas. But the spiritual genius of Fr. McGivney went in a different direction. He chose the principles of charity, unity, and fraternity.

These principles direct a Knight of Columbus not toward himself, but away from himself to the service of others. With these principles, Fr. McGivney put into concrete and practical form a basic truth of the Catholic faith: a person finds fulfillment for himself by serving others. In this very practical way, Fr. McGivney, by founding the Knights of Columbus, gave a very concrete answer to the questions: What is an adequate understanding of the human person, and what is it that gives human life meaning and purpose?

In the three preparatory years leading up to the Jubilee Year 2000, which focused upon the Father, the Son, and the Holy Spirit, Pope John Paul II also answered those questions in a very precise way. He reminded us that the dignity, freedom, and destiny of each person are understandable only in the light of the life of Jesus Christ. Yet, the life of the Son is understandable only in the light of his relationship with the Father. Thus the harmonious divine communion of the Father, Son, and Holy Spirit provides the basis and form for communion between human persons. As John Paul II has observed, all human relationships find their ultimate source, form, and meaning in the relationship among the Divine Persons of the Holy Trinity.

Thus the trinitarian structure of human communion provides the only adequate foundation upon which to build a new "Culture of Life." This is in part the reason why Pope John Paul II has placed such emphasis upon "solidarity" as a fundamental human relationship.

And this is why the Knights of Columbus are today so committed to building a new "Culture of Life." Fr. McGivney's vision of charity, unity, and fraternity has provided for more than a century an answer to the question: "What is man, O Lord?" In doing so, he has led the Knights of Columbus to enter more fully into an associational and ecclesial communion rooted in an inherently trinitarian structure.

As we enter the new millennium, an age surely to be defined in many ways by advances in biotechnology, we look forward to working with you in the pursuit of a new Culture of Life.

Carl A. Anderson
Supreme Knight
Knights of Columbus

The Human Person

ON THE HORIZON IN BIOTECHNOLOGY

WILLIAM B. HURLBUT, M.D.

Fifty years ago Aldous Huxley, anticipating the transformation of human life through advances in biology as the "final and most searching revolution," asserted that "this really revolutionary revolution is to be achieved, not in the external world, but in the souls and flesh of human beings."[1] Today, with the new powers of biomedical technology, we are on the early edge of that radical revision of human existence. We are experiencing its immediate practical dilemmas, but more significantly, its challenges to our understanding of the human person and our place within the natural order. So we must ask ourselves: In the most profound sense, what is human place and human purpose?

If we stand back and look at the earth and our place within the cosmos, we are immediately struck by what an extraordinary reality we are part of. Consider how human existence is located between infinities, between the infinitely large and the infinitely small, the vast realms of cosmic space and the tini-

---

[1] Aldous Huxley, *Brave New World* (New York: Harper and Row Publishers, 1932), xi–xii.

11

est subatomic particles. Each of us is fashioned in the silence of the womb from the most minute molecules—atomic assemblies one millionth of a hair's width, forming proteins, membranes, organelles and cells, cells forming organs and organ systems—a fantastic symphony of process. And all this is played out on the small stage of the earth within a universe so vast that the number of stars actually exceeds the number of grains of sand on all the beaches of the world. There are two hundred billion stars in our galaxy alone, stretching a million trillion miles, spiraling through space, and one hundred billion galaxies as large—galaxies and galaxy clusters in an almost unimaginable array. And now cosmologists are saying that our universe may be just one bubble in a larger froth of universes!

Consider also how we are placed between the infinities of time—the very fast and the very slow: the frenzied dance of atoms colliding a billion times a second; enzymes that convert substrate at a million a minute; nerve networks and synapses of such millisecond speed as to integrate a hundred trillion neuronal connections in a fantastic behind-the-scenes choreography that makes possible the thought and movement of real time, life at the level of human perception. Yet the rhythm of the days and the cycles of the seasons is made possible by the enormous stretches of cosmic time—processes so gradual as to appear from a human perspective as a changeless backdrop for the unfolding of human history. The ancient earth, prepared through thousands of years of geologic upheaval, revolves around our slowly aging sun. The sun circles once in three hundred million years around the center of our galaxy. Our galaxy spirals through billions of years in the outrush of the expanding universe. Human life, fragilely balanced, is situated between the tiniest microseconds and the enormous eons of cosmic time.

And consider one further set of infinities—how we are located within the infinite range of the possible. If the forces that hold the atomic nucleus together were only a few percent stronger or weaker, the diversity of elements could not exist. If our sun were just slightly more or less luminous, life could not have emerged on the earth. Think of the absolutely amazing sequence of events that allowed the increasing complexity of evolving life forms. It is as though this universe had us in mind from its origins. Physicist Joel Primack says of the basic laws of physics, "this appears to be the only set of principles and conditions that could produce conscious life."[2]

---

[2]Conversation with author, Stanford, California, October 2000.

All these circumstances of size and time and fragile complexity converge to make possible human existence. What is obvious through the study of biology is that as the universe has unfolded, human life has been formed and fashioned by the forces of the earth and intricately interwoven with the whole physical and biological world, like a hand made to fit an existing glove. The forces of nature have defined our lives and formed our being. From the earth we draw forth our human place and human purpose. Indeed, this is the meaning of the word "human": its Latin root comes from the word for earth or soil. We are the "creature of the earth."

Try for a moment to imagine the "morning of the world" where we lived in intimate unity with the earth, where our lives and their meaning were embedded in the forces that shaped our being—guided, nourished, and constrained by the earth that formed us. If these early ages of humanity were anything like the hunter-gatherers of today, we can get some sense of this primary relationship with nature. The physician-anthropologist Melvin Konner spent two years living among the !Kung in South Africa. He wrote:

> The !Kung are frequently ill; their physical complaints cannot be lost on any anthropologist who spends more than a few days among them. They suffer from endemic subclinical infectious disease processes from malaria to gastrointestinal infections to tuberculosis and many others.... The !Kung appear to escape the diseases of civilization"—ulcers and high blood pressure, for instance—but there are plenty of other diseases they do not escape.[3]

He goes on to say,

> The !Kung are not satisfied with their lot ... they are neither at peace with or inured to the many losses those bleak mortality curves deliver to them; and ... they are more or less continuously envious of people who are better off, both within and outside their own society.[4]

Clearly, this is not Rousseau's Noble Savage. Rather there is disease and dissatisfaction, a restless longing—pointing to an awareness of something beyond the horizons of life. Nonetheless, Konner goes on to describe the meaningful realities of their lives: joys and generosities, community and sharing, and sensitivity to sickness and old age.

---

[3]Melvin Konner, *The Tangled Wing: Biological Constraints on the Human Spirit* (New York: Henry Holt Company, 1982), 372.

[4]Ibid., 374.

So it must have been for the small beginnings of humanity: embedded in the cycles and seasons of the earth; intimately interwoven within the fabric of the natural world; framed and constrained by the forces of biological nature; deeply connected by the meaningful relationship between body and being. But how did we get from tiny huts on the windswept savanna to the vast skyline of New York City? Yet for all the drama of this transformation, it is perhaps less significant than our current leap into the world of biotechnology, because this is not a transformation of the external world, but of our very selves.

Genetic Screening

Consider the technology of preimplantation genetic diagnosis where a cell is removed from an eight-cell embryo, created by in vitro fertilization (IVF), before the embryo is implanted into the womb of the prospective mother. This screening process was first performed in September 1992 (ancient history by the current pace of biotechnology), and led to the successful delivery of a healthy baby girl. At the time one commentator described it as "an early glitch on the radar screen of the twenty-first century."[5] This procedure opened a small doorway into a vast realm of medical manipulation, of technological intervention at the most basic level of human life. It represents not just a starting point for the radical revision of our physical process, but also, perhaps, a more radical revision of our sense of who we are.

In this procedure, hormonal treatments are used to induce the maturation of an unnaturally large number of eggs, a dozen or more. These are located by sonogram and removed by needle aspiration. They are then mixed with donated sperm and incubated in the laboratory three to four days to the eight-cell stage. A single cell is removed (others will grow in and compensate) and the cell's DNA, which weighs just six billionths of a gram, is extracted and amplified to get a sample large enough for genetic analysis. In the case mentioned above, both parents were known to carry the gene for cystic fibrosis. Though they were each genetically recessive, carrying just a single copy of the gene, without the screening their child would have had a twenty-five percent chance of receiving the gene from each parent and therefore of manifesting the disease.

---

[5]Gina Kolata, "Genetic Defects Detected in Embryo Just Days Old," *New York Times*, September 24, 1992, A1.

One can sympathize with the parents' desire to have a healthy child, but this procedure raises fundamental questions concerning the idea of quality controls—who should decide and by what criteria of value, and is the making of such decisions a proper role for medicine? Francis Collins, who discovered the gene for cystic fibrosis, was a practicing clinician before his current job as Director of the Human Genome Project. I asked him whether he had ever taken care of a patient with cystic fibrosis who wished he had never been born. He said that he had not.

Furthermore, with preimplantation genetic selection, where is the ethical stopping point? The several thousand diseases and disorders with known genetic causes cover a wide range of human conditions. Will we filter out only serious and untreatable diseases, or unattractive and socially undesirable conditions as well? Deafness, dwarfism, dyslexia, albinism, asthma—who decides? We now have 750 genetic tests, but the completion of the Human Genome Project will soon bring a greater understanding of the genetic basis of human variation in even complex traits such as proclivity toward mental disease, dimensions of intelligence, and possibly sexual orientation. What is the role for parental preference in selecting for height, hair and eye color, and possibly basic personality profile?

The Nobel laureate Walter Gilbert has said that by the year 2010 he expects every newborn child to go home from the hospital with a CD containing the information of his of her entire genome.[6] Recent studies, however, seem to indicate that in the near future, most meaningful human variation may be detected for less than one hundred dollars using a simple postage-stamp-size DNA chip.[7] But these tests will not be absolutely predictive. Because of the complexity of genetic interactions and environmental influences, genetic tests can give only a statistical probability for the expression or severity of a phenotypic trait. Genetics is not as deterministic as many people believe. Even in identical twins, diseases such as cystic fibrosis can have different manifestations.

Nevertheless, whether it be at the eight-cell stage or during gestation in the hidden recesses of the womb, genetic testing is already leading to what has been described as "micro-

---

[6]See [mbb.harvard.edu/undergrad/amdn/courses/seminars.html].

[7]Gordon Ringold, CEO of Affymax, conversation with author, Stanford, California, October 1999.

eugenics": the one-at-a-time, parentally-chosen prenatal "quality control" of prospective offspring. The driving force behind such practice is as deep as human prejudice and social imperative. This will be especially true with mounting pressure to limit family size for population control; if you are allowed only one child, you want it to be the perfect child. In China, where a whole generation will not know the natural meaning of the terms brother or sister, aunt, uncle, or cousin, the preference for male children has resulted in a differential male to female birth rate of five to four. This is happening not only in urban centers. Throughout India where a similar cultural bias holds, there are so called "fertility clinics" where, in ninety-nine percent of the procedures, females are aborted.[8] A similar danger may apply in the United States. A survey showed that one percent of Americans favor gender-selection abortion, six percent favor abortion for a fetus carrying genes for developing Alzheimer disease after age sixty-five, and eleven percent favor prenatal screening followed by abortion if the baby is shown to have a proclivity toward obesity.[9]

The problems associated with genetic screening will be the first great ethical dilemma resulting from the sequencing and analysis of the human genome. Long before any prospects for altering individual fates or engineering human evolution, we will have the capacity to screen and select, to quantify, calibrate, and control human genetic variation. Even if one takes a moral stance against abortion, we will still face the dilemma of what has been called "toxic knowledge" (do we really want to know so much about ourselves and each other?). Having information about our genetic endowments may lead to destructive preoccupation and worry, intolerance, and social discrimination. Already there is evidence that we may be able to predict, at least on a statistical basis, educational potential and proclivities for antisocial behavior.[10] What values will guide us as we consider preemptive medical interventions during gestation, or prejudicial selection for employment or educational opportunities? In cases of prenatal selection for parental prefer-

---

[8]Neera Sohoni, *Burden of Girlhood: A Global Inquiry into the Status of Girls* (Oakland, CA: Third Party Publishing Company, 1995).

[9]Francis Collins, "Isolating Disease Genes: Diagnosing and Understanding the Nature of Disease," paper given at the Stanford Centennial Symposium Conference, 1991.

[10]Peter McGuffin, Brien Riley, and Robert Plomin, "Toward Behavioral Genomics," *Science* 291.5507 (February 16, 2001): 1232–1249.

ences, what burden of expectation is placed upon the child that would not be present if reproduction were approached as an open and unconditional commitment of love?

In March of 1999, an ad was run in the *Stanford Daily*, the Stanford University student newspaper, offering fifty thousand dollars for the donation of eggs for IVF. The ad specified donor qualifications of five feet ten inches in height and a 1400 score on the SATs, representing the top one percent for physical height and intellectual aptitude. The lawyer representing the prospective parents said they were a tall, intelligent, athletic couple. However, a student who was a finalist in the selection process told me that after filling out nearly thirty pages of personal, social and medical information, she was introduced to the prospective mother—who turned out to be five feet, one inch. I suspect they wanted a son, and studies indicate that taller men do have distinct social and economic advantages.

Already one percent of births in the U.S. are the result of artificial insemination by donor (AID).[11] We now have pre-fertilization sex selection and internet mail-order sperm banks; one California firm has shipped sperm to over thirty countries. There is a sperm bank that offers insemination with semen from Nobel Prize winners. Last year a website appeared in which purchasers were invited to bid on eggs from female fashion models. One major teaching hospital keeps a catalogue, complete with donor profiles, of spare frozen embryos available for adoption.

Even where this technology is limited to infertile couples using their own gametes, the dilemma of excess embryos is both poignant and problematic. At the Stanford Assisted Reproductive Technologies Clinic there are several canisters of liquid nitrogen cradling over fourteen hundred human embryos kept for possible future implantation.[12] (This must be the highest population density in human history.) The problem, of course, is their ambiguous status as human persons. At least two countries, Spain and England, have laws mandating the destruction of frozen embryos after a set period of years. But on a private level the issues are more complicated. In one case, twin embryos were implanted and brought to term in two different pregnancies seven-

---

[11]See "Donor Insemination: Information and Support," [www.angelfire.com/bc/donorinsemination/] (December 9, 2002).

[12]Dr. Barry Bahr, Director of SARTC, conversation with author, Stanford, California, January 2001.

and-a-half years apart.[13] In other cases there have been custody battles over frozen embryos after divorces, and even a dispute over inheritance when a wealthy couple died in an airplane crash and left several frozen embryonic heirs.

Stem Cell Technology

Most recently the status of the human embryo has become the subject of further controversy because of advances in stem cell technology. This new area of scientific inquiry, emerging from a convergence of advances in genetics, cytology, and developmental biology, holds the promise of becoming the most revolutionary therapeutic tool in the history of medicine. Throughout development there is a progressive differentiation into specific tissues and organs, but early in this process certain embryonic cells retain the power to form a wide variety of specialized cell types. These pluripotent embryonic stem cells can now be removed from a four-to-five-day-old embryo and multiplied indefinitely in laboratory culture. With our increasing knowledge of genetics we are gaining an understanding of the molecular signals and growth factors that coax these cells down specific pathways of differentiation. With this knowledge may come the ability to grow an unlimited supply of cells and tissues (and possibly whole organs such as livers or hearts) for medical therapy. The possibilities implied by this are truly fantastic. Cell therapy has been proposed for the treatment of a wide array of severe and intractable medical conditions, ranging from Parkinson disease and Alzheimer's, to liver failure, diabetes, and retinal degeneration. Furthermore, replacement cells have been demonstrated to migrate and seamlessly integrate into proper position and function within such sites as injured neural tissue and compromised cardiac muscle.[14] Neural stem cells show promise in the treatment of spinal cord injury, stroke, and multiple sclerosis, and have been proposed as a possible restorative therapy for mental retardation and brain damage from head trauma.

In addition, these embryonic stem cells can be genetically modified in controlled laboratory culture and serve as the vehicle for gene therapy when transplanted to specific sites within the body. Through new techniques of intrauterine surgery, such

---

[13]*Stanford Daily*, February 18, 1998.

[14]Ronald McKay, "Stem Cells—Hype and Hope," *Nature* 406.6794 (July 27, 2000): 363.

cell therapies could be used to treat developmental disorders or preclude later expression of genetic disease. Indeed, at least in mice, it has been demonstrated that genetically modified embryonic stem cells can be infused into an early embryo and will integrate and differentiate into a wide array of developing tissues.[15] This may one day lead to germ line gene therapy for the elimination of genetic disorders such as Huntington disease or Tay Sachs disease. However, these techniques could also be used in attempts to extend or enhance human functioning or in a larger project of technologically guided human evolution.

More immediately, however, there are ethical dilemmas associated with the procurement and the generative potential of embryonic stem cells. The current source of these cells is from extra embryos from fertility clinics. Though these embryos are already destined for destruction, obtaining the stem cells necessitates the termination of a life in process.

The ambiguity concerning the status and dignity of the embryo reaches a deeper dilemma with embryo splitting and cloning. As mentioned earlier, it is possible to remove a cell from an early embryo and the other cells will compensate. In fact the removed cell can be placed on a parallel petri dish and will itself develop into a complete embryo. A similar process occurs naturally and is the source of identical twins and triplets. Some have suggested that, during IVF, identical embryos be intentionally created by such embryo splitting and stored as repositories of "spare parts." As needed, the twin could be thawed, gestated, and harvested for immunologically compatible embryonic stem cells or fetal tissues for therapeutic transplantation. This possibility, however, may not require the forethought and cost of maintaining a frozen identical twin. Recent success in producing animal clones through the technique of nuclear transfer suggests the likely success of human cloning. Here the nucleus from an adult cell, such as a skin cell, is fused with an egg cell which has had its nucleus removed. Remarkably, the adult nucleus appears to be restored to its primitive undifferentiated status and in some cases becomes capable of directing the development of a complete new embryo.[16]

---

[15]Gretchen Vogel, "Mice Cloned from Cultured Stem Cells," *Science* 286.5499 (December 24, 1999): 2437.

[16]In fact, in none of the cases studied up to this point has there been true genetic identity between the original organism and the clone, as a result of complexities associated with methylation, etc.

The idea of using this technology for human reproductive cloning has been greeted with nearly universal abhorrence, though one couple has provided five hundred thousand dollars to an offshore company called Clonaid to produce a genetic replica from frozen cells of their deceased infant daughter.[17] One can feel their weight of grief while not approving the solution to their sorrow. But, even if we prohibit reproductive cloning, some argue that this technology offers the perfect solution for the production of abundant immune compatible tissues for transplantation as mentioned above. This technique, now called "therapeutic cloning" or "cloning for biomedical research,"[18] is now legal in several countries and would allow the production of an unlimited supply of patient-specific tissues.

Cloning for biomedical research would involve the intentional creation and nurturing of an embryo which would then be sacrificed to harvest the needed cells, tissues, or organs. The argument is made that if abortion is legal, that is, if a developing life can be terminated with no reason given, then why not for a good reason? One must admit there is a certain perverse logic to this argument.

There may, however, be ways to produce specific embryonic cells that lack the potential to form full embryos. Searching out the moral meaning of "partial generative potential" or "parts apart from wholes" will require both solid scientific knowledge and ethical discernment. Notwithstanding the abhorrent images and emotional atmospherics evoked by stem cell technology, its positive possibilities for medical benefit make it imperative that the Church, while forthrightly defending human dignity, search out with subtlety of discernment every morally permissible means to draw on this tremendous therapeutic potential.

One very hopeful development in this direction comes from recent discoveries of the extraordinary, and largely unanticipated, presence and plasticity of adult stem cells. These cells serve throughout life as a source of regenerative potential to renew and repair body tissues. Astonishingly, it now appears that a single stem cell can restore the trillions of cells of the

---

[17]Constance Holden, "Company Gets Funds to Clone Baby," *Science* 289.5488 (September 29, 2000): 2271.

[18]This term has been adopted by the President's Council on Bioethics in an effort to more accurately describe the action it represents.

entire blood system.[19] It is obvious that such systems as skin and blood must have a regenerative source but, to date, more than thirty major types of adult stem cells have been identified including bone, cartilage, liver, pancreas, and remarkably, brain tissues. Most surprising, and contrary to a long standing doctrine of developmental biology, these cells appear to be able, with specific molecular signals, to abandon their differentiated identities and convert to other tissue types or possibly even revert to the open potential of pluripotent embryonic stem cells. Cells from bone marrow, muscle, and brain, for example, have each been coaxed to form cells with morphologic and phenotypic characteristics of endothelium, neurons, glia, and hepatocytes.[20]

Promise and Peril

These discoveries, at the convergence of developmental biology and molecular genetics, open up a wide range of tools for scientific inquiry and therapeutic intervention. With the completion of the Human Genome Project, we have the DNA sequence for the entire complement of human genes, estimated to number about forty thousand.[21] These genes code for proteins, the major structural and functional molecular components of the body. In the emerging field of Functional Genomics, new tools of analysis allow us to monitor gene expression and thereby the production and function of the various proteins. Studies in Proteomics, the identification and characterization of each protein and its structure, will let us

---

[19]Irving Weissman, M.D., conversation with author, Stanford, California, March 2000. This research has been done with mice, but very likely indicates a similar potential in human hematopoietic stem cells.

[20]Y. Jiang et al., "Multipotent progenitor cells can be isolated from postnatal murine bone marrow, muscle, and brain," *Experimental Hematology* 30.8 (August 2002): 896–904. Notwithstanding the hopeful implications of these early reports, one must be cautious in making scientific claims concerning transdifferentiation. Events such as fusion, contamination, and genetic alteration may confuse the research results.

[21]Earlier estimates had placed the number at one hundred thousand. Prior reports based on the partially completed sequence of the human genome revised this figure down to around thirty-five thousand. More recently the estimated number of genes is again rising. This uncertainty concerning such a basic issue is an important indicator of how little we know about human genetics.

map out the metabolic and developmental processes of both health and disease. They will also allow us to identify potential protein targets for pharmacologic interventions.[22] It would be difficult to overstate the implications of this for understanding and intervening in human life. To date all of our therapeutic drugs target a mere four hundred protein sites. Within ten years scientists anticipate the number of identified protein targets to increase to ten thousand, greatly extending our ability to treat disease and alter basic physiological and psychological functioning.[23]

Together with stem cell technologies, this understanding of molecular biology will allow the testing of new drugs within controlled cell cultures and greatly accelerate the pace of discovery and clinical application. Hundreds of new biotechnology firms with a capitalization of nearly a half a trillion dollars are using systems such as automated combinatorial chemistry to generate and screen millions of compounds in the search for biologically active agents.[24] New animal models for studying human disease processes are being created by hybridizing human genes and human cells within laboratory-engineered mice, rats, and rabbits. Sheep and pigs have been genetically engineered as "living factories" to produce large quantities of human proteins in their milk or semen for use as pharmaceuticals. Vast banks of tissues and even entire organs may be grown in laboratories or within bioengineered animals for transplantation into humans. Edible vaccines are being developed using genetically modified plants, making possible the eradication of many pathogens and parasites such as guinea worm disease and polio. One after another diseases that have marred the face of humanity for thousands of years will be cured or effectively controlled. New understanding of the neurosciences will extend these discoveries into treatments of mental disease and then, perhaps, into enhancement of normal cognitive functioning. Moreover, cell therapies with targeted growth factors will allow regeneration and repair and then, possibly, rejuvenation of body parts and technological extensions of the human lifespan.

---

[22]Carol Ezzell, "Beyond the Human Genome," *Scientific American* 283.1 (July 2000): 64–72.

[23]Ibid., 66.

[24]Editorial, "The Promise of Proteomics," *Nature* 402.6763 (December 16, 1999): 703.

Notwithstanding their many potential benefits, these technologies are greatly altering our relationship with the natural world, our social realities, and our sense of personal identity and destiny. Along with our new powers, a new meaning of medicine is emerging, and with it a dramatic transformation in our understanding of human place and human purpose. This change in the use of our scientific knowledge has already been building along the gradients of our appetites, ambitions, and open aspirations. Slowly but steadily our technology is eroding the boundaries and balance of natural human life and alienating us from the larger motions of our biological and spiritual significance.

Beyond Healing

An ad for Rogaine, a pharmaceutical agent used to stimulate hair growth, states, "*If you're losing your hair you no longer have a reason to lose hope.... See your doctor.*" This exemplifies the "medicalization" of natural human life. When I was a medical student, male pattern baldness was not considered a disease. The traditional role of medicine has been to cure disease and alleviate suffering. Seen in this way the task of the physician was to restore health so the patient could return to the process of living a normal life. This idea was put succinctly by the Roman physician Galen when he said, "The physician is only nature's assistant." Now, however, with our new biomedical technology, all that is changing. The idea of healing is being displaced by a broader metaphor of "liberation"—freedom. This is not just freedom from sickness, but from distress, struggle, and even the constraints of natural life processes— freedom from all that is unattractive, imperfect, or just inconvenient. In short, medical science is being used as a tool in a project of technological transformation in the pursuit of happiness and the quest for human perfection.

Over the past few decades there has been a dramatic rise in medical interventions for cosmetic purposes. Liposuction is now the second most common surgical procedure (second only to abortion). In the U.S. two million women have had breast augmentations and surveys show that 5–13% of adolescent boys now use steroids, not just for athletic performance, but for a stronger, more masculine appearance. Furthermore, according to an editorial in the *Journal of the American Medical Association*, over half the use of synthetic human growth hormone (hGH) is driven not by medical indications, but by "a cultural 'heightism' that permeates Ameri-

23

can society."[25] Indeed, taller men are rated more attractive and make higher salaries. These practices amount to medical treatment for self-confidence and social advantage.

Human beings have always sought to control their bodies and shape their identities, but now, freed from natural constraints, there is an ever more radical revision of human flesh, of form and function. One after another, our technology is shattering the boundaries that have defined our lives and framed our self-understanding. What nature has joined together technology is putting asunder: contraception—freedom to have sexual intercourse without producing a baby; AID and IVF—freedom to have a baby without sexual union—a practice now common among lesbian couples; postmenopausal pregnancy, now possible with egg donation—freedom to have a baby when it is career-convenient.

The contraceptive pill was a major milestone in the transformation of the role of medicine and in our whole attitude toward natural life processes. For the first time tens of millions of women were using a medicine to "cure" something that is not a disease. But do we understand the personal, social, and spiritual implications of such reordered dynamics of life? The sexual revolution resulted in increased promiscuity, sexually transmitted disease, and infertility. Last year there were three million new cases of chlamydia[26] (the major cause of infertility and subsequent IVF). For many women contraception has meant postponement of childbearing into years of decreased fertility and the heartache of an empty cradle. One sorrowful woman writes of the "unused magic" of her body.[27] Nonetheless, clinical trials are now being conducted on a chemical contraceptive that puts menstruation on a trimonthly cycle. There is even talk of rendering menstruation "obsolete" by hormonal manipulation.[28]

---

[25]Leona Cuttler et al., "Short stature and growth hormone therapy: a national study of physician recommendation patterns, *Journal of the American Medical Association* 276.7 (August 21, 1996): 531–537.

[26]See [www.cdc.gov/nchstp/dstd/Fact_Sheets/chlamydia_facts .html].

[27]Anne Taylor Fleming, *Motherhood Deferred: A Woman's Journey* (New York: Fawcett Columbine, 1994), 16.

[28]Rachel Sobel, "Is a Monthly Period Still Necessary? A New Pill Questions Basics of Female Biology," *U.S. News and World Report*, June 5, 2000, 58.

With our new knowledge of biology and the precision tools of our vastly expanded pharmacopia, many other aspects of human life from fetal development to aging will become the focus of medical intervention and revision. As a society we seem to have an eagerness, even an expectation, for a medical solution to nearly every type of personal or social problem. Nowhere is the potential for both medical benefit and disruption of natural life processes greater than in neurological functioning. New tools for imaging and monitoring brain function, together with a deepening understanding of brain chemistry, are setting the stage for a wide range of interventions in human behavior. Already, in what has been called "cosmetic psychopharmacology," drugs such as Prozac and Paxil, designed to treat serious depression, are being used by the "worried well" to improve the sense of buoyancy and self-esteem. One psychiatrist suggests that the late 1990s stock-market "irrational exuberance" might be explained in part by the chemical inhibition of natural caution related to the ten million prescriptions for Prozac![29]

On a more serious front is the medicalization of restlessness in preadolescent boys; according to one study more than fifty percent of children treated with Ritalin failed to meet even the broadest definition of attention deficit hyperactivity disorder.[30] This raises a major objection against the "medicalization" of life—it can be a false and deceptive solution, a "quick fix" by a society unwilling to address problems at their personal and social roots. This is leading us toward one the most difficult ethical dilemmas of all: the biological foundations of freedom and responsibility, and the "medicalization" of criminal behavior. Recent studies suggest that genetic and developmental factors underlie some forms of antisocial behavior including addiction, impulsive aggression, and aberrant sexual compulsions.[31] Will we now turn to such solutions as preemptive pharmacologic therapy in childhood, mandatory electrode implants, or psychosurgery as an alternative to adult impris-

---

[29]Randolf M. Nesse, "Is the Market on Prozac?" *Edge* 64 (February 28, 2000), [http://www.edge.org/documents/archive/edge64.html] (December 9, 2002).

[30]Eliot Marshall, "Duke Study Faults Overuse of Stimulants for Children," *Science* 289.5480 (August 4, 2000): 721.

[31]Kenneth Blum et al., "Reward Deficiency Syndrome," *American Scientist* 84.2 (March–April 1996): 132–143.

onment? These possibilities would be potential extensions of current research and therapies. Surgical techniques that involve the focal destruction of brain tissue are already being used therapeutically with some success in the treatment of obsessive-compulsive disorder and Parkinson disease, and electrical implants are now used to control tremors and intractable pain.

Technological Transcendence

Furthermore, consider the possibilities as we reach deeper into the neurophysiology of behavior. Likely areas for biological intervention include early education, athletic performance, and even romantic love. Already there are drugs to increase memory, aggressive confidence, and sexual desire. What strange realms will we enter as we unravel the foundational substrata of perception and consciousness, as we gain power in the production and management of all pains and pleasures? Some say this is good, that it will bring new forms of recreation, religious experience, and self-exploration. Others say that it is human destiny to express ourselves beyond the limits of nature, that it was evolution's plan that man should become the artist of his own creation. They call for a major project of biological reengineering of humanity, of self-guided evolutionary extension of the human species. Among the proposals for the implementation of such technological transcendence are varied forms of "cyborgs": human-machine hybrids. Research to create neurologically integrated artificial prostheses for amputees and paraplegics has led to the development of computer interfaces that can form direct functional connections with nerve fibers. Some scientists propose that in the future we will all carry implantable brain chips that will extend the dynamic range of our senses, enhance memory and information access, and allow constant communication within a wide social network.

Others have in mind a more direct biological transformation through genetic engineering and a radical revision of both mind and body. They see the human species as a mere stepping-stone in a longer and larger evolutionary project. Their proposals for "Posthumans," superior by design, include new modes of perception, increased intelligence, and physical immortality; and sometimes a new world order of love, peace, and harmony brought about by biological redesign of the human mind.

There is a great deal of extravagant speculation and journalistic exaggeration concerning the bioengineered revision of the human race. Indeed, this seems to be one of the foremost products of biotechnology. It is used to sell magazines, but it is also given assent by an astonishing number of serious scientists. In just two examples from an issue of *Science* magazine, the physicist Stephen Hawking says "It is likely that we will be able to completely redesign [the human genome] in the next one thousand [years]," and William Haseltine, head of Human Genome Sciences, marking the creation of the Society of Regenerative Medicine, states, "The real goal is to keep people alive forever."[32] (Talk about irrational exuberance!)

We've been through something like this before in the first half of the twentieth century. In the words of Hermann Muller, a 1946 Nobel laureate, programs of planned eugenics "provide the opportunity to guide human evolution to make unlimited progress in the genetic constitution of man, to match and reinforce his cultural progress, and reciprocally, to be reinforced by it in a never ending succession."[33] Muller, at the time an ardent supporter of the communist movement, suggested Marx and Lenin be our models of ideal humanity. Later, when he became disillusioned with communism he changed his suggested models to Pasteur and Lincoln. This "flavor of the month" problem underscores the dilemma—if we are to improve on natural humanity, what will be our standards and values? But more to the point, the biology of engineered evolution would most likely exceed not just our wisdom but also our technical control.

Until now, we have had very few clear and successful applications of gene therapy, though most scientists working in the field say that in another ten to fifteen years we will be treating some specific disorders. But repairing genes, like fixing a link in a broken chain, will be far simpler than implementing a project of eugenic engineering. Genes are not Legos, those little children's blocks that can be stuck together in random combinations and constructions. Most genes affect many traits, and most traits are effected by many genes. Genes code for proteins that in turn interact within the body to influence numerous traits, so a benefit in one direction may produce a deficit in another. Notwithstanding a *Time* magazine cover, irresponsibly implying we may soon increase intelligence by

---

[32]"Futurology Corner," *Science* 290.5500 (December 22, 2000): 2249.

[33]Charles Frankel, "The Specter of Eugenics," *Commentary* 57.3 (March 1974): 31.

manipulating "the I.Q. gene,"[34] most of what we really care about (overall beauty or intelligence or longevity) will be so complicated and so deeply influenced by developmental and environmental factors outside our control that they will be beyond the reach of projects of genetic technology. One might say humanity will be protected from its own folly by the sheer complexity of the problem.

Even if we could reconstruct ourselves to emphasize particular characteristics of mind or body, the diversity of our species is a benefit, not a defect. Furthermore, our overall design as a species is that of a "general purpose organism." We have adapted for adaptability, not for a narrow specialization. Our very strength is in creative flexibility, freedom, and open indeterminacy. These are what give us our extraordinary capacities: our comprehensive consciousness and controlling powers. Our species may already *be* the end of the arrow of evolution, the optimal design for fullest functioning and flourishing of life. Indeed, it is our very strength that is now threatening us. Liberated from the immediacies of mere survival, we are opened to imagination, to the ambitions of technological self-transformation that could shatter the fragile balance of our physical and psychological functioning. Even without radical revisions, the more mundane "medicalizations" of natural life processes could disrupt the fine-tuning of our freedom and the relational dynamics of meaningful human life.

Voluntary Poverty

As embodied beings, embedded in the ecology of the earth, we are precariously poised between a desperate flight from disease and disorder, and the drives and desires of our open aspirations. But natural desires are motivations—directions not destinations. Placed within the constraints of nature, our positive passions are used in the service of survival and the full flourishing of life. Freed from these limitations by advancing technology, our desires become for us a danger—the persuasions and preoccupations of pleasure and pride.

As we replace landscape with machinescape, we are severed from the necessities and self-evident truths of natural life. The testimony of transcendent significance within the drama of human existence, of love and death, is replaced by a sense of futility. The original radiance and vitality of the cosmos, its

---

[34]*Time* 154.11 (September 13, 1999).

order and beauty, are obscured by the conviction that all of living nature is mere matter and information to be reshuffled and reassigned for the projects of the human will. On every front from egg donation to obesity to aging, we set in motion an ever-escalating imperative of perfection, a technological standard that displaces the natural with an artificial ideal. Even our children become not our progeny, but our products. Torn between arrogance and anxious striving, we become caught in a self-sustaining conflict of conscience fueled by a shifting valence of vulnerability and vanity, of insecurity and intolerance. Heroic struggle and sacrifice give way to a competitive rivalry corrosive to human self-concept and community; and love gives way to the lust of luxury.

It has been said that there is no cure for luxury, and that luxury is a disease that we have that kills other people. Indeed, the United Nations reports that thirty thousand children under the age of five die every day from starvation and diseases related to malnutrition. Will we now turn and use the great power of our advancing biotechnology in the pursuit of vanity? Will we all become its victims?

My thoughts keep coming back to little Saint Francis. Perhaps no man ever walked the world with a greater joy for the beauty of life. He is well known for his appreciation of the natural order and his communion with its creatures. What is less known is the story of his courageous self-surrender, his willing abnegation of his privileged life of prosperity, and pleasure. He freely gave up his wealth and position, and turned to care for the wretched, repulsive bodies of the outcast lepers, and there he found the resonant reaches of the fullness of love.

The great French theologian Louis Bouyer wrote, "Man can recover true life and preserve the cosmos only by rediscovering that a certain voluntary poverty is the condition for possessing the world in a way that will not reduce it to ashes."[35] It is an irony of history that the word "humility" shares with the word "human" the Latin root meaning earth or soil. Brought forth from the earth, between infinities, we are the "creatures of the earth" and we should be humble within it. What great possibilities for good lie ahead with our advancing biotechnology! Received with reverence and used in a spirit of humility and love, these extraordinary powers may become instruments of praise for the glory of God.

---

[35]Louis Bouyer, *Cosmos* (Peterham, MA: St. Bede's Publications, 1982), 160.

"WHAT IS MAN?" THE MYSTERY BEYOND THE GENOME

EDMUND D. PELLEGRINO, M.D.

*"What is man that you should be mindful of him,
or the son of man that you should care for him?"*
Psalm 8[1]

The Psalmist's question has always intrigued man. Mystics and theologians have long sought the answer in man's soul, while philosophers looked to his physical nature. More recently, psychologists have turned to man's unconscious self and biologists to his body. Physicians have tried in vain to find the soul somewhere in their anatomical dissections. Yet, the question persists. Unless we know who we are, and what we are, we cannot know how we should live with neighbors, our God, and ourselves. Without an understanding of man, there is no grounding for ethics.

It is precisely for this reason that the image and idea of man that we draw from the Human Genome Project (HGP) is so important for ethics in general and for the ethics of the Ge-

---

[1]*New American Bible* (New York: P.J. Kennedy, 1970). Throughout this essay I will use the word *man* in its generic and classical sense to indicate a member of the species *Homo sapiens* without regard to gender, race, religion, etc. I take this to be the source of its use in Psalm 8.

nome Project itself. If, as some seem convinced, man is completely explicable in terms of his genetic constitution, then ethics is merely a product of evolutionary biology, a device for protecting the gene pool. If, however, man is more than his genes, then we must look beyond genetics and neurobiology for our moral guidelines, and the search of traditional moral philosophy for truth remains valid.

The thesis I will argue here is that the new knowledge about man deriving from the HGP does not support the assertion of enthusiastic materialists for a deterministic physiochemical explanation of the mystery of man. Indeed, an assessment of what is known thus far only deepens the mystery. It challenges our metaphysical compass points even while we further explore the intricate and complex workings of the human organism.

For Christians, man is God's creation formed in his image and likeness, so loved by God that God himself became a man. The mystery of man is linked to the mystery of the Incarnation, as well as the ontology of embodiment of a spiritual soul in a material body. Though we know not the full mystery, we are grateful indeed that God "cares" for mortal man. We desire to know all we can about man. But we also know we shall never "explain" him, because, as St. Paul says, "in Christians, the fullness of Deity resides in bodily form. Yours is a share of this fullness in Him who is the head of every spirituality and power" (Col. 2:9–10). Or, as Karl Rahner puts it so pithily, "Man is the question to which there is no answer."[2]

For the nonbeliever, however, the mystery of man is just another problem to be solved, another secret of nature to be unraveled and mastered to suit man's purposes. For the nonbeliever, the mystery must somehow be reduced to man's material substance—to his atoms (Democritus [460–457 B.C.] and Lucretius [94–55 B.C.]), to the machinery of his body (la Mettrie [A.D. 1709–1751])[3], to evolutionary selection (Darwin, A.D. 1809–1882), or to the fortuitous interplay of energy and force fields. Each "solution" exalts some observable facet of man's material body to explain away the mystery of his mind, soul, and being. But, as each new "explanation" is explored more fully, the mystery only deepens.

---

[2]Karl Rahner, *Christians at the Crossroads* (New York: Seabury. 1975), 11.

[3]Julien Offray de la Mettrie, *L'homme machine* (Princeton: Princeton University Press, 1960).

The latest, and assuredly the most powerful, attempt to answer the question of man, the reduction of the mystery of man to his genetic constitution, to a molecular entity is, as of yet, difficult to define precisely. The Human Genome Project, by "mapping" the genes located on all forty-six human chromosomes, has expanded our knowledge of man's biology immensely. Whether it has helped us to "understand" man any better is more than dubious.

This is not to depreciate the potential uses of knowledge of the human genome, e.g., the eradication of hereditary diseases, prolongation of life, enhancement of health, better agriculture and animal husbandry, etc. More problematic are the predictions of genetic control of human behavior and of human evolution, or the engineering of a "better" human species. To date, the evident therapeutic possibilities are still largely possibilities and not actualities. Today's obsession with perfect health or immortality through genetic engineering has nonetheless verged on "genomania."[4]

The Human Genome Project is neither a salvation theme to lead us out of the affliction of original sin, nor is it a device of the devil to undermine religion or God's creation. Both extremes are avoidable if we restrain the temptation to genomania and the geneticization of all human life on the one hand, and the rejection of genetic science on the other.[5] Like every new explosion of human knowledge, genetic science and molecular biology must be examined critically and subjected to careful ethical, philosophical, and theological criticism.

Such an inquiry must start with an examination of the state of the art and of the conceptual and practical problems already evident in the knowledge derived from the Human Genome Project. There is a multitude of issues to be resolved if that knowledge is to be used wisely and well. Exaggerated hopes and unrestrained utopianism have dominated too much of the early discourse. Thus far, metaphysical pronouncements by some of the scientists who have pioneered the field are simply unwarranted extrapolations of their own presuppositions. Simi-

---

[4]Ruth Hubbard, "Genomania and Health," *American Scientist* 83.1 (January–February 1995): 8–10.

[5]John M. Opitz, "Afterword: The Geneticization of Western Civilization: Blessing or Bane?" in *Controlling Our Destinies: Historical, Philosophical, Ethical, and Theological Perspectives on the Human Genome Project*, ed. Philip R. Sloan (Notre Dame, IN: University of Notre Dame Press, 2000), 429–450.

larly, pious denunciations without examination of the scientific data are equally unwarranted.

What is clear is that we are at another historical moment of profound challenge to our worldview and our notion of human nature, equivalent to the Copernican and the Darwinian revolutions. We will know much more than we do now of the complexity and wonder of the human organism. What we learn will give us further confirmation of the unfathomable mystery and presence of God in the creation of man and the cosmos. When all is said and done, the mystery of man will remain.

The Human Genome Project

This is not the place to attempt a summary of the Human Genome Project (HGP). Four excellent compilations have appeared recently, and much of what follows has been drawn from these sources.[6] A quick sketch of the scientific study of inheritance does seem in order to set the background against which it should be clear that the mystery of man is, and will remain, beyond the era of the Human Genome.

In 1866, in an Augustinian monastery, the monk Gregor Mendel studied the inheritance of certain traits in peas. He developed the hypothesis that these traits were transmitted to offspring by their parents in a predictable way through the agency of certain factors present in the biological constitution of these parents. In the years following, patterns of inheritance were studied in a wide variety of species, from fruit flies to humans. Thirty-five years after Mendel's discovery, Walter Sutton further hypothesized that Mendel's factors were not only found in humans, but were the basis of inheritance and were located on chromosomes in each human cell. In all, the fundamental unit of heredity was shown to be the gene, a physical unit for transmission of specifications of form and function from one generation to the next. In 1953, Watson and Crick determined that the gene was a segment of double helical DNA, containing the information needed to synthesize specific proteins within

---

[6]Barbara R. Jasny and Donald Kennedy, "The Human Genome," *Science* 291.5507 (February 16, 2001): 1153–1306; Opinion, "The Human Genome," *Nature* 409.6822 (February 15, 2001): 745–768; "Opportunities for Medical Research," *Journal of the American Medical Association* 285.5 (entire issue February 7, 2001): 491–686; Francis S. Collins and Victor A. McKusick, "Implications of the Human Genome Project for Medical Science," *Journal of the American Medical Association* 285.5 (February 7, 2001): 540–545.

the cell. Every chromosome in every cell is encoded with this information, which expresses itself in complex and incompletely understood ways. The sum total of this genetic information is the "genome," a specific pattern, unique for each individual and each species of living things.

The Human Genome Project is an international collaborative effort to precisely locate each of the approximately thirty thousand genes on each of the forty-six chromosomes of the human cell. These genes, and the three billion or more base pairs[7] that extend into the DNA-helix-like ladder rungs, contain the information necessary to synthesize all the proteins in the one hundred trillion cells of the human body. The genome is the sum of the genetic DNA sequences on all the chromosomes in each human cell. It locates, in precise order, the sequence in the strings of nucleic acids and the rungs of the DNA ladder. The proteins produced by these genes shape the anatomy of the human body and its functions. But when missing or altered, these proteins can produce disease and point to its possible treatment.

As the recent reports indicate, this process of "mapping" the human genome is virtually complete. The next phase of interpretation must focus on how the map works, on what it means and, in essence, where it leads. The human genome "map" is not the end of a quest, it is the beginning. In a lucid commentary, two of the leaders in the field, geneticists Francis Collins and Victor McKusick, list some of the next logical steps in understanding the genome, e.g., sequencing the genomes of other organisms, and comparing and contrasting the differences; developing a catalogue of variations and clues for estimating genetic medical risks; understanding more about gene expression and alterations of expression in disease, particularly in polygenic disease; interactions between genes; interactions between genes and environment, etc.[8] Other questions to be investigated include the regulation and the functional significance of the concomitant, so-called junk DNA.

Clearly, the knowledge already in hand, and that to be added in the years immediately ahead, will raise many concerns about how, and under what conditions, the knowledge of the Genome

---

[7]See R.J. Trent, "A Nucleotide Base with Its Complementary Base" in *Molecular Medicine: An Introductory Text*, 2d ed. (New York: Churchill Livingstone, 1997).

[8]See Collins and McKusick, "Implications of the Human Genome Project."

will be used. What about genetic privacy and ownership of individual genome maps? Should screening be voluntary or involuntary? Are patenting and commercialization justifiable? What diseases should be treated and who should determine who should be treated? What limitations should be placed on enhancement versus therapeutic genetic engineering? Should we genetically alter babies according to parents' requests? Should we apply genetic information to eugenics? Are there experiments which ought not be done?

Our answers to these ethical questions, whether implicit or explicit, are grounded in the philosophical or theological anthropology we espouse. The central ethical question is whether the Human Genome Project tells us anything significant about the nature of man well enough to alter the way we justify our ethical decisions regarding the use of knowledge from the HGP. The remainder of this essay will focus on the uncertainties in the physiological meanings of the HGP and on the fallacy of deriving a conception of human nature grounded solely in man's genetic constitution alone. In effect, do the Human Genome Project results justify a purely biological anthropology and, by inference, a biologically based ethic?

The Concept of the Gene

Let us look, first, at some of the empirical and theoretical uncertainties in genetics and the Human Genome Project and then at its limitations as evidence for a reductionist and materialist understanding of man and ethics.

It is remarkable how an entity whose precise definition is still debated among scientists has been so empowered, reified, and personified that it has been labeled "selfish"[9] and seemingly granted a will of its own. Francis Crick, co-discoverer of the structure of DNA, sees in the gene the realization of the hope of every materialist for the explanation of all life in terms of physics and chemistry.[10] Though the terms are different, Crick's hope is another resurrection of the ancient atomistic materialism of Democritus and Lucretius. A much more defensible and moderate view is offered by Francis

---

[9]Richard Dawkins, *The Selfish Gene* (New York: Oxford University Press, 1976).

[10]Francis Crick, *Of Molecules and Men* (Seattle: University of Washington Press, 1966); Francis Crick, *The Astonishing Hypothesis*, (New York: Simon & Schuster, Touchstone Books, 1995).

Collins,[11] the director of the project, and Theodosius Dobzhansky,[12] one of the earlier prophets of modern genetics.

Crick's assurance, however, is belied by the history of the concept itself.[13] There is no doubt that DNA is a fact, a definable chemical entity satisfying the criteria of scientific evidence. But the concept of the "gene" is far from being such a definable, structured entity. This uncertainty derives in part from the newer knowledge of nucleotide sequencing coming out of the Genome Project itself. It is now clear that genes may have many structures, locations, and functions. Distinctions between and among its putative biochemical, biological, and embryological properties have yet to be clarified. The biological primacy of the gene has yet to earn the status of a settled question among biologists.[14]

None of this is to question the value of the gene concept as an organizing or heuristic notion, or to seek its abandonment, as some have suggested. For the moment, a working definition of the gene like that of Singer and Berg's "combination of DNA segments that together constitute an expressible unit, expression leading to the formation of one or more specific functional gene products"[15] seems both useful and tenable.

This definition allows flexibility in defining location, chemical composition, structure, function, and causal efficacy of genes while avoiding the ideological implications of earlier definitions which the Genome Project itself has shown to be imprecise and still changing. From the point of view of those of us who are not experts in molecular genetics, these uncertainties of definition serve as a warning against the exaggerated claims of deterministic omnipotence attributed by some who have made genes out to be "mystical" entities.

[11]Francis S. Collins, "Medical and Societal Consequences of the Human Genome Project," *New England Journal of Medicine* 341.1 (July 1, 1999): 28–36.

[12]Theodosius Dobzhansky, "Changing Man, Modern Evolutionary Biology Justifies an Optimistic View of Man's Biological Future," *Science* 155 (January 27, 1967): 409–414.

[13]Evelyn Fox Keller, "Is There an Organism in This Text?" in *Controlling our Destinies*, ed. Sloan, 273–289.

[14]Evelyn Fox Keller, *The Century of the Gene* (Cambridge: Harvard University Press, 2000).

[15]M. Singer and P. Berg, *Genes and Genomes: A Changing Perspective* (Mill Valley, CA: University Science Books, 1991), 622.

Similar questions now engage biologists, geneticists, and embryologists with respect to how much control genes actually, in fact, exert in embryogenesis, disease causation, and morphogenesis. Accumulating evidence suggests that it is just as likely that the organism, as a whole, exerts at least as much control over the gene as the gene exerts over the whole organism.[16] This is evident in the changing viewpoints about genetic explanations of disease, cure, and the relationship of genes to behavior and human cognition.

Genes and Disease

Some of the most exciting and hopeful information coming out of the Human Genome Project is the evidence of a genetic component in hereditable human diseases. To the layman, it often seems that the location of a gene responsible for some human disease is uncovered daily. The prototype concept that this suggests is the notion that for every disease there is one abnormal gene responsible. Sooner or later, it is reasoned, gene therapy tuned to this mutation will cure or prevent this disease. By manufacturing, replicating, substituting, or modifying the identified gene, it is supposed, inherited disease can be cured, prevented, or eliminated.

This one-gene, one-disease idea gained support in 1949 when sickle cell anemia was shown to be a single defect in the hemoglobin molecule. Similarly, genes for Huntington disease, cystic fibrosis, several forms of breast and colon cancer, certain types of muscular dystrophy, and Alzheimer disease have also been "located." As a result, hope arose for sufferers and their families that cure was at hand. Some investigators, popularizers, and biotechnology companies did little to dampen the excessive enthusiasm. Some enhanced the possibilities to garner support for more research or to provide investment opportunities. Public enthusiasm and the lure of profit make perfect soil for unrestrained optimistic speculation.

More careful analyses have shown that the single-gene, single-disease concept is an oversimplification of the complex interactions surrounding the action of genes in an organism as intricate as the human body. It is becoming clearer that possession of a genetic defect, associated with a particular disease in a particular person, only indicates the possibility and probability of its expression, not the certitude. Acknowledgment

---

[16]Elizabeth Pennisi, "Behind the Scenes of Gene Expression," *Science* 293.5532 (August 10, 2001): 1064–1067.

of this fact would spare patients some of the anxieties of ill-considered screening or exaggerated hope for gene therapy. Public access to genetic information on the Internet makes access to reliable knowledge and qualified genetic counselors mandatory. If false hopes are not to proliferate, responsible reporting and interpretation are ethical obligations at every level of genetic science and application.

Humans must come to appreciate that all of us carry some genetic mutations. The expression of each mutation, however, is the sum total of many interacting factors about which knowledge is now incomplete. The relationship of the "external" macroenvironment and of the "internal" microenvironment is still largely a question. The same is true of the interactions with other facilitating or inhibiting genes, associated DNA and the existence or nonexistence of final common pathways where the actions of different genes may converge.

The single-gene paradigm is being transformed as the HGP provides new knowledge about the interactions of genes with other physiological entities at the level of the cell and the organism. Taking its place is a multigene paradigm of disease in which different phenotypes can emerge depending on chromosomal background.[17] Polygenic origin seems to be the rule for some of the more common and important chronic diseases like diabetes mellitus, coronary artery disease, or stroke.[18] Here, the pathways to reliable detection, prediction, prevention, and treatment are long and arduous. These are, in a sense, destinations to which the genome map alone cannot take us.

Of course, the search for genetic markers and mutants should continue even if the first seemingly "responsible" gene turns out to behave in more complex ways than originally supposed. An economy of pretension about what single genes can affect should not deter investigators from closer analysis of phenotypes, or from the study of the relations of genotypes through use of more traditional methods of genetic study in clinical situations.

As these relations are pursued, it seems clearer that the way the information transmitted by the gene is actually effected

---

[17]S. M. Kiesewetter et al., "A Mutation in CFTR Produces Different Phenotypes Depending on Chromosomal Background," *Nature Genetics* 5 (1993): 274–278.

[18]Collins and McKusick, "Implications of the Human Genome Project."

at the molecular level is still tentative. For example, the folding and activity of proteins are yet to be fully related to the sequence of nucleotides and nucleic acids.[19] Also, molecular evolutionary changes may not follow Mendelian principles or Darwinian selection. Instead, they may result from random drift in the fixation of alleles.[20] The determinism some would assign to the human organism and its higher functions is yet to be confirmed at the molecular level from whence that determinism is thought to arise. The disastrous results of our ignorance of these factors is evident in the high morbidity and mortality rate of cloned animals. Simple possession of a gene is, clearly, no guarantee of its expression in a clone.[21]

What all of this tells us is that, even at the level of empirical evidence, our understanding of the mechanisms of gene action is far from certain. Current theories of information transfer, the process of evolution, and the mechanism for turning genes "on" and "off" are poorly comprehended. Surely, this information should be pursued scientifically. But when there is so much yet to learn about man's material constitution and function, it seems beyond any scientifically defensible justification to conclude that man is "nothing but" the product of his genes.

These difficulties are not resolved by some of the newer concepts and efforts like biological holism and emergence. The theory of emergence postulates that the phenotype is not a device of the gene to replicate itself because the elements of a biological system interact in unpredictable ways. It is not the mere sum of its genetic components. A holistic biology has recently been proposed to explain the emergence of complexity and to counter the classical notions of interactions of the genotype, phenotype, and environment. A new synthesis has been advanced to unite the animate and the inanimate.[22]

---

[19]Hubert P. Yookey, *Information Theory and Molecular Biology* (Cambridge: Cambridge University Press, 1992).

[20]Moto o Kimura, "The Neutral Theory of Evolution: A Review of Recent Evidence," *Japanese Journal of Genetics* 66 (1991): 367–38; Francisco Ayala, "Neutralism and Selectionism: The Molecular Clock," *Gene* 261.1 (December 30, 2000): 27–33.

[21]Rick Weiss, "Human Cloning Bid Stirs Experts' Anger," *Washington Post*, March 7, 2001, A1, A6.

[22]J. Scott Turner, *The Extended Organism: The Physiology of Animal-Built Structures* (Cambridge: Harvard University Press, 2000).

As a matter of fact, the more carefully and critically one examines the biochemical foundations and the deterministic and reductionistic theories of molecular biology, the more one must struggle to avoid the scientific evidence for design.[23] Indeed, it seems that the materialist must struggle hard to avoid the presence of a designer for such an intricate and finely balanced system as the human cell, to say nothing of the mind, brain, soul, and psyche.

Genes and Behavior

Matters become vastly more complicated when we attempt to connect genes, gene mutations, and the behavior of animals. Most behavioral genetics has been done in relatively simple animals like crayfish, round worms, or mollusks. Transferring these data to human behavior is a leap across multiple levels of complexity that challenge credibility, to say the least. From the single neuron of a round worm to the cognitive network of the human cerebral cortex is a jump beyond the present powers of molecular genetics or deductive logic.

Schaffner has carefully and precisely outlined eight rules relating genes to behavior which underscore the dangers of any simplistic description of either animal or human neural behavior.[24] Schaffner shows how complex the relationships are between genes and neurons, neurons and behavior, embryogenesis and neural connections, and genes and gene-environment interactions. These complexities should constrain deductions from genetic studies about the genesis of Alzheimer disease, bipolar psychoses, or schizophrenia, for example. However, they should not preclude further research into the possibility of understanding and better treating these diseases.

Far beyond any single gene or polygenic hypothesis at present is the relationship of gene action to such "higher" human behavioral phenomena as imagination, artistic creativity, formation of abstract ideas, the capacity to deal with novelty or to devise new solutions to puzzling problems and ideas. For example, in the genetics of perfect pitch, one cannot avoid the suggestions of hereditability of musical talent in the Bach

---

[23]Michael J. Behe, *Darwin's Black Box: The Biochemical Challenge to Evolution* (New York: The Free Press, 1996).

[24]Kenneth F. Schaffner, "Genes, Behavior, and Developmental Emergentism: One Process, Indivisible?" *Philosophy of Science* 65.2 (June 1998): 209–252.

family, for example. But no set of data or hypothesis has explained how the molecular structure of DNA can produce complicated productions of human creativity like a poem, a sculpture, or musical masterpieces.[25] No nature-nurture or heredity-environment interaction explains the "Well-Tempered Clavier."

Like computer algorithms, genetic algorithms fall short of explaining the defining characteristics of human intelligence—language, speech recognition, common sense, emotional expression, or personality. Chess-playing computers like "Deep Blue" are tests of processing power, not of intelligence. This is true of computer-generated music, architecture, and art. In the end, the computer is programmed by the human mind even if the computer can later be taught to "learn" by experience, to change its own programs, or to "invent" new ones. Where genes fit in the longstanding debates about mind-body, mind-soul relationships is not discernible using present data. Human spirituality, emotions, and personality will not be found on any genetic map.

Genes and Neurosciences

An even more powerful challenge to our traditional ideas about human nature is the merging of the HGP with the newer knowledge of brain science. The 1990s were dubbed the "decade of the brain." While not as far along in detail as the HGP, discoveries about the structure and function of the human brain are prodigious. They are, for some, a further attempt to explain the mystery of man in biophysical and biochemical terms.

Brain research is construed by some to provide evidence for seeing the brain as nothing but a very sophisticated computer. Algorithms for artificial intelligence are expected to resolve the mysteries of human intellection. Sooner or later it is predicted that the human brain will be surpassed by one of its own creations—a self-replicating supercomputer. Accordingly, intelligence, mind, soul, and imagination, it is concluded, are simply epiphenomena, by-products of the physical workings of brain cells without any causal power of their own.

When genetic science is added to brain science, the functions of brain, mind, and soul, as well, are made to be geneti-

---

[25]Stephen Jay Gould, "Message from a Mouse: The Error of Speaking of 'Genes' for This or That Behavior," *Time* 154.10 (September 13, 1999): 42.

cally "hardwired" into the neurons. Imagination, memory, moods, reason, judgment, consciousness, and conscience are predetermined by our genomes—the unique configuration of genetic information which identifies us as biological individuals. What is asserted is that the thirty thousand plus genes of the human body are sufficient to explain the content and connections of thousands of millions of neurons and the almost infinite number of connections between and among them.

Some proponents of the Human Genome Project are asking us to believe the embodiment of the human soul is simply the result of the "wiring" of our brains and no longer a mystery. Are spiritual experiences then located in the left parietal lobe because it "lights up" on Positron Emission Tomography (PET) when patients have religious experiences? Is there a "God module" in the left parietal lobe?[26]

Similarly, is possession of the XYY chromosome the origin of aggression? Is there a genetically determined location in the brain for alcoholism, substance abuse, or sexual orientation? What does this mean for such human realities as guilt, accountability, responsibility?[27] What does this mean for our doctrines of sin, atonement, and redemption? The implications of a concept of human nature based in a combination of genetics and neurobiology are all truly stupendous.

Clearly, the demonstration that the brain is functioning and that one part of it relates to certain functions from emotions to spiritual experience neither explains nor reduces these experiences to matter. Indeed, it indicates the close union of matter and form, of soul and body, an affirmation, not a denial of the Thomistic and scholastic idea of hylomorphism. If there were no synergy between spirit and matter, soul and brain, then Cartesian dualism could be inferred as easily as monistic materialism. Current findings with the PET scan are no more effective in negating hylomorphism than the more naïve attacks of the seventeenth century critics like John Locke and Robert Boyle.

Genetics and brain science, either alone or in combination, do not suffice to resolve the mystery of man. They do, how-

[26]V.S. Ramachandran and Sandra Blakeslee, *Phantoms in the Brain: Probing the Mysteries of the Human Mind* (New York: William Morrow, 1998).

[27]Frank H. Marsh and Janet Katz, ed., *Biology, Crime and Ethics: A Study of Biological Explanations of Criminal Behavior* (Cincinnati: Anderson Publishing Co., 1985).

ever, both solve the "problem" of mind-brain and soul-body relationships. Not only do they deepen the mystery, they increase our wonderment at the enormous complexity of God's creative entry into the formation and sustenance of the being He created in his own Image.

Any answer to the mystery of man requires more than a biochemical, biophysical account of the human body. It must account for mind and soul as well. Body is the vehicle for contact with the material world. Mind is whereby we encounter the immaterial, the world of the abstract ideas like time, number, truth, freedom, creativity, right and wrong conduct, and good and evil. Soul is that whereby we seek for an encounter with the Divine, the Transcendent, and to have religious experiences. Their domains of experience may differ, but body, mind, and spirit are marvelously one, in a mysterious way beyond mere gene expression.

That is why ethics cannot be "wired" by genetics and evolution into the nervous system, and why it cannot be defined as simply a technique to keep the genetic material intact. Such an ethic is based in a biological anthropology. A major challenge for Catholic, Christian, and other religious ethical systems will be to counter this kind of biologism. We must be careful not to imbibe sociobiology with our admiration for the scientific accomplishments of the HGP.

Scientists who assert that man is "nothing but" molecular interactions, genetic DNA, or mysterious energy force fields, are interposing their own scientistic ideologies into the interpretation of sometimes very scanty data. These ideologies are no less ideological because they are advanced by credible scientists. We can only ask that scientists apply the same rigor to their extrascientific extrapolations as they do to the data of their colleagues. If truth is one, and if truth cannot contradict truth, then scientific methodology must recognize its own limitations when it encounters phenomena which are not reducible to matter and energy.

The Lure of "Final" Theories

For some time now, biologists have envied the perceived precision of theoretical physics.[28] In their zeal for reductionis-

---

[28]Steven Weinberg, *Dreams of a Final Theory: The Scientist's Search for the Ultimate Laws of Nature* (New York: Vintage Books, 1994).

tic theories of human nature, they hope to approximate the unified theories of the workings of the physical universe, presumably dispelling the mysteries of the cosmos and of God. A look at the history of such attempts should have a chastening effect on the biologist's enthusiasm.

One need start only with the nineteenth century. Shortly after James Clerk Maxwell described the laws of electromagnetism, many thought that only the details were left to be discovered. They failed then, as they do now, to reckon the appearance of the quantum and relativity theories. Much has been learned, to be sure, but the reduction of the workings of the universe to "four" forces or six "numbers" only poses new conundrums.[29]

Thus, it seems at the level of molecular biology, and at the level of the physical universe, reductionism only leads to more mystery. St. Thomas's pronouncement, "*Omnia exeunt in Mysterium*" [all ends in mystery], seems applicable to biology as it does to physics and metaphysics, as well.

Again, none of this is to ridicule, or discourage the search for unifying theories. Each attempt brings us a little closer to understanding how things work, and can confer new powers on human beings. What the limitations of reductionistic theories do teach us, however, is the need for logical humility and parsimony in extrapolation from realms of science to realms of metaphysics.

The Human Genome Project: Where Has It Taken Us?

What I have tried to do in this essay is to assess, from a nonexpert physician's point of view, some of the limits the general public must put on the exaggerated claims of those engaged in genetic research. As I have already stressed, this in no way implies a depreciation of the Genome Project, of genetic research, or of attempts to understand as much as we can about human existence and the human body. Rather, the Human Genome Project is an astounding accomplishment of scientific know-how and international collaboration. Its promise for diagnosing, understanding, preventing, and treating human and animal disease is truly enormous and beyond our present capacities to envision fully. It is no exaggeration to place the Ge-

[29]Martin Rees, *Just Six Numbers: The Deep Forces That Shape the Universe* (New York: Basic Books, 2000).

nome Project in the same category of accomplishment and possibility as Galilean, Newtonian, and quantum physics, and the theory of relativity, the idea of Darwinian evolution, Freudian psychology, and present-day astrophysics and cosmology.

Each of these infusions of new knowledge has challenged us to reexamine the inherited views of ourselves and of nature. Each has the potential to exaggerate some facet of man or the natural world and to give the illusion of an answer to the mystery of *bios*, *theos*, or cosmos. Each, therefore, has potential for good and harm.

But it is the task of ethics to define the constraints of good and bad, right and wrong uses of knowledge. Happily, the Human Genome Project has recognized this need from its inception and has subsequently apportioned some of its funds for ethical research. Unfortunately, to date, most of the ethical opinion has been driven by the principle of utility. Believers in Christian ethics must exert equal effort to stand against utility as the guiding principle and to insist that no matter how much "good" may emerge from a technical advance, the "good" cannot be ethically derived from a morally dubious interpretation of the results of the HGP.

Completion of the mapping of the thirty thousand plus genes and some three billion base pairs in the human chromosomes is a beginning, not the end of our understanding of molecular genetics. The Genome Project has given us a map, but as of yet, it is a two-dimensional map. It tells us where things are, but provides no instruction on how they are regulated to interact with each other in either normal functioning or in disease. We still need to know much more about the terrain, the physiological climate, the levels of organization, and the ecology and environment within which the genes function in the map. We do not know yet what other maps may have to be superimposed to correct our wrong impressions, clarify relationships, and provide the level of detail we need to use and understand the information gained from the Human Genome Project.

For the Catholic Christian, there are legitimate and illegitimate areas for study of the gene.[30] Scientific methodologies and goals for the Christian must be responsive to normative guidance from sources not controllable by science: revelation, Church teaching, tradition, etc. This fact does not prohibit the

---

[30]J.D. Cassidy, O.P., and Edmund D. Pellegrino, "A Catholic Perspective on Human Gene Therapy," *International Journal of Bioethics* 4.1 (March 7, 1993): 11–17.

search for truth, but it does require that even the search for truth respond to moral guidance.

The challenge for Christians, as scientists, is to participate in genetic and genome research. Their task is to demonstrate that scientifically valid and usable knowledge can be obtained via morally licit methods of experimentation. Human embryos, for example, cannot be acceptably destroyed to advance knowledge or even to elaborate treatments; human beings cannot be mere objects of research despite their so-called "persistent vegetative state"; potential good for the many cannot override the dignity and worth of even the least among us.

Roman Catholic biologists and chemists should take participation in genetic research as a high priority. So too must Catholic colleges, universities, and foundations. Their search must be for the use of knowledge in ways that do not violate Christian anthropology. It is now clear, for example, that stem cell research can be pursued with adult cells from many sources and need not involve the death of embryos. Gene therapy for specific diseases needs to be pursued but should be restrained by ethical concern for experimental subjects, despite the impetus of pride and profit that can endanger them.

Cloning of other humans is at the pinnacle of human hubris and without redeeming features. Cloning as an experimental or therapeutic procedure confined to specific cells and tissues is licit and important to pursue. This is not the place to enlarge on research areas open to Catholic and Christian molecular biologists and geneticists. Rather, I wish only to urge them and their institutions to engage this field of research. We need to devise ethically licit ways of gaining and using the knowledge, not only of the Genome Project, but of the multiple pathways to which it is the gateway.

The challenge to Catholic theologians and philosophers is to engage the metaphysical claims of the reductionists and the moral philosophy they engender. They must draw on the rich tradition of natural law, bring it into dialogue with biological anthropology, and inform themselves about what is, and is not, demonstrated as valid genetic research. In the end, it is the task of Catholic Christian anthropology, to provide a convincing response and critique of the dangers of a purely biological ethic.[31]

---

[31]Germain Grisez, "On Religion: Bioethics and Christian Anthropology," *National Catholic Bioethics Quarterly* 1.1 (Spring 2001): 33–40; Kevin Fitzgerald, "Ethical Analysis of Human Genetic Interven-

Pursued with humility and scientific rigor, the Human Genome Project is neither a salvation theme nor a replay of Faust's compact with the devil. Rather, it has the potential to open an era of good for humans and a deeper understanding of God's hand in the creation of man. Beyond the genome, man will remain a mystery, the question without an answer.

Blaise Pascal, no mean scientist himself, put it well: "God alone can teach us that knowledge of our own nature, which of ourselves we cannot have."[32]

---

tions: A Response from Karl Rahner to the Need for Integrating Philosophical Anthropologies and Current Scientific Knowledge" (Ph.D. diss., Georgetown University, 2000).

[32]Blaise Pascal, *Pensées*, trans. H.F. Stewart (New York: Pantheon, 1950), 141.

SOUL AND THE
TRANSCENDENCE OF THE
HUMAN PERSON

ROBERT SOKOLOWSKI

The Problem

As human beings we are animal and organic, but we also carry out spiritual activities. We are not only animals; there is a spiritual side to us that becomes manifest in what we do. The spiritual activities of human beings stem from our reason and the kind of freedom that reason makes possible.

What do we mean when we say that human beings have a spiritual dimension? We mean that in some of our activities we go beyond or transcend material conditions. We go beyond the restrictions of space and time, and the kind of causality that is proper to material things. We do things that cannot be explained materially. To speak adequately about ourselves, we must use categories different from those used to speak about matter.

It seems intuitively obvious that human beings enjoy such a spiritual dimension in their lives, that we are more than material and animal beings. In our contemporary culture, however, to claim that we have a spiritual dimension is very controversial, because much of our culture takes it for granted that we are simply material things. It assumes that anything that seems spiritual will sooner or later be explained away as the working out of material bodies and forces. Nothing spiritual has yet, in fact, been explained away in that manner, but the culture assumes that it inevitably will be, that everything spiritual will be boiled down to the material.[1]

Such a reduction of the spiritual to the material, such a denial of the spiritual, is carried out in three lines of argument. First, our rational activities, both our knowing and our willing, are said to be reducible to neurological processes; the mind and the will are to be reduced to the brain and nervous system. The mental and spiritual dimensions of man are reduced to biology. The neuroscientist William H. Calvin says, for example, at the end of one of his analyses of the brain, "None of this explains how the neurons accomplish these functions ...

---

[1]The recent much discussed book by Edward O. Wilson, *Consilience: The Unity of Knowledge* (New York: Knopf, 1998), attempts to give a comprehensive explanation of the human person and society in terms of the evolution of matter; human beings are said to be "organic machines" (82). A typical passage is the following: "As late as 1970 most scientists thought the concept of mind a topic best left to philosophers. Now the issue has been joined where it belongs, at the juncture of biology and psychology. With the aid of powerful new techniques, researchers have shifted the frame of discourse to a new way of thinking, expressed in the language of nerve cells, neurotransmitters, hormone surges, and recurrent neural networks. The cutting edge of the endeavor is cognitive neuroscience ...." (99). Three other widely noted books expressing a reductive understanding of man are: Francis Crick, *The Astonishing Hypothesis: The Scientific Search for the Soul* (New York: Charles Scribner's Sons, 1994); Paul M. Churchland, *The Engine of Reason, the Seat of the Soul: A Philosophical Journey into the Brain* (Cambridge, MA: The MIT Press, 1995); and Patricia Smith Churchland, *Neurophilosophy: Toward a Unified Science of the Mind-Brain* (Cambridge, MA: The MIT Press, 1986). There are many works being published now about the brain and human activity. The works of the neuroscientist Antonio Damasio are especially interesting philosophically and less reductive; see *Descartes' Error: Emotion, Reason, and the Human Brain* (New York: G. P. Putnam's Sons, 1994), and *The Feeling of What Happens: Body and Emotion in the Making of Consciousness* (New York: Harcourt Brace, 1999). For a less deterministic understanding of the function of genes, see Evelyn Fox Keller, *Reconfiguring Life: Metaphors of Twentieth-Century Biology* (New York: Columbia University Press, 1995), and *The Century of the Gene* (Cambridge, MA: Harvard University Press, 2000).

but I hope that the foregoing explains why brain researchers expect to find the mind in the brain."[2]

Second, our neurological processes and all the rest of our bodily activities are explained as the effects of molecular biology; they all come from the chemical structure of cells and the activity of the DNA working in the cells of our bodies. The biology of our bodies is reduced to the chemistry and physics of our cells. The logic, therefore, is that the spiritual is reduced to the biological, and the biological to the chemical.

The third line of argument against the spiritual dimension of man lies in the Darwinian theory of evolution, which thinks it can give us the complete story of how we developed from physics to chemistry to biology to psychology. Evolution as an ideology is very important in the cultural controversies of our present day because it claims that someday it will be able to show how the specifically human being developed randomly from matter and material forces. It claims it will show that the more complex, the organic, and the spiritual are the resultants of material forces combining with random mutations. They are the resultants of a combination of necessity and chance, with no providential or creative intelligence behind them, and no transcendence of space, time, and material causality. Darwinian evolution is presented as the alternative to the biblical narrative.

The extent to which evolution is related to rejection of divine providence is candidly expressed in the following passage by the linguist Derek Bickerton. He complains that many behavioral scientists want to deny that human beings are different from other animal species. He disagrees with them, and says that their failure to recognize the obvious difference is a case of throwing the baby out with the bathwater. He explains:

> The bathwater that such thinkers wish to throw out is the Judeo-Christian belief that humans are different from and superior to other creatures because God created them separately and divinely ordained that they should be so. The baby in this case is the lamentable fact that, cut it how you will, we *are* radically different from other species, and to deny it (while eating a microwaved dinner or riding in a mass-produced auto) is hypocrisy.[3]

---

[2]William H. Calvin and Derek Bickerton, *Lingua ex Machina: Reconciling Darwin and Chomsky with the Human Brain* (Cambridge, MA: The MIT Press, 2000), 72.

[3]See Derek Bickerton, *Language and Human Behavior* (Seattle: University of Washington Press, 1995), 113.

Bickerton takes it for granted that belief in creation should be eliminated, but he wants to preserve the difference between man and animals, a difference that was introduced through the clever calculation of evolution:

> We are different, not because God specially made us so, not even because it is better so—it is perfectly possible to believe we are different and to wish we were not. We are different because evolution made us that way, because (for perfectly valid evolutionary reasons) it provided us, but no other species, with language.[4]

This reductionist and materialist world picture has not in fact been proved by science, but it is taken for granted as the way things must be. It is a hope, and those who believe in it claim that sooner or later science will substantiate this picture. If Darwinian evolution is taken as a replacement for the biblical narrative, this hope in the ultimate explanatory triumph of science may be taken as the replacement for biblical hope in salvation.

I dare say that many Christians, who believe in a spiritual dimension of man and the world, may themselves be unduly disturbed by this picture of things. They may worry that the materialists might really know something, or may find something out, that will definitively show that what seem to be spiritual activities are really only the outcome of material forces. In this cultural situation, the challenge for Christian philosophy and theology is to bring out the distinctive nature of spiritual activities, to show what they are and how they are present to us. Our challenge is to show that human beings are truly involved in activities that transcend the restrictions of space, time, and matter, that we do accomplish things that are spiritual, and that therefore we are spiritual as well as material beings. It is to show that what we mean by spirit could never be boiled down to matter, that it would be meaningless and incoherent to try to reduce the spiritual in this way.

Spirit and Soul

I wish to take some steps in meeting this challenge, and I will begin by making a distinction between spirit and soul. Soul is proper to all material living things: men, animals, insects, and plants have soul, but soul is not something separate from their bodies. Soul is not a separate entity, not a ghost in a machine, and we must try to speak about it in such a way that we

---

[4]Ibid.

do not give the impression that it is a separate thing. A good synonym for the word *soul* is *animation*, and this word has an important advantage: we are much less tempted to think that the animation of a living thing could be found apart from that thing. You cannot have animation all by itself; it has to animate something. Animation or soul makes a living thing to be one thing, one entity. Soul is the unity of a living thing. It is also the source of the activities that thing carries out when it acts as a unified whole. Soul becomes visible not by introspection, but publicly in the conduct of living things; and the kind of soul a living thing has is shown by the kind of activities the living thing can perform.

In animals and plants, soul is exhausted in animating the body. In the case of man, however, there is an aspect of soul that enables him to live a spiritual life. In man the one soul is the source of a spiritual life as well as a bodily, organic one. Spirit and soul are not simply equivalent.[5]

I would like to spell out more fully the distinction between spirit and soul. Consider angels. Angels are spirits, but they are not souls and do not have soul. They live a spiritual life; they think and they decide, but they do not animate a body. By their nature, they are not involved in matter. But let us turn our attention away from the high domain of angelic spirits; let us look at more familiar worldly things, where traces of spirit without soul can also be found, in bodily things that show the effect of human rational activity. Consider something like furniture. Furniture shows the effect of human reason, and therefore it has something spiritual about it; it has a residue of spirit, but it does not have soul because the life of reason that generates furniture does not dwell in the wood itself; it dwells in the human beings who make the furniture. In man, however, the same principle that generates spiritual activities also enlivens the body.

---

[5]A very good treatment of the difference between soul and spirit can be found in Dom Anscar Vonier, O.S.B., "The Human Soul," in *The Collected Works of Abbot Vonier*, vol. 3 (Westminster, MD: The Newman Press, 1952), 3–66. Abbot Vonier wrote this work on the soul in 1912. Some of the best contemporary philosophical writing about the human person and the human spirit can be found in the work of the Catholic philosopher from Munich Robert Spaemann. See his *Personen: Versuche über den Unterschied zwischen Etwas und Jemand* (Stuttgart: Klett-Cotta, 1996).

In both angels and furniture we have spirit but no soul, and in animals and plants we have soul but no spirit. Only in man do we have both soul and spirit; we have the animation that makes a body into one organic, active entity, but we also have the capacity to act in ways that are not limited to the body, ways that transcend the space, time, and causation that are proper to the body. This combination of matter, soul, and spirit makes man a great mystery. We are bodies and animals, and yet we live a spiritual life. It is only because we are both spiritual and ensouled that we form churches and political societies, establish universities and research centers and libraries, perform dramas and watch football games, and come together as we do here and now to talk about what we are.

Before we leave this distinction between spirit and soul, let us say a word about the way soul is related to the body that it animates. In living things, it is not true that all the causation comes from the material elements in the body. It is not true that a living thing is merely the effect of the mechanical and electromagnetic forces that combine to make it up; rather, in living things the matter itself is shaped and reshaped by the thing as a whole, and hence by the animation of the thing. All forms of soul—plant, animal, and human—change the matter that they enliven. Living things contain chemical molecules that are not found and could not be found except in living things, and matter is formulated into these incredibly complex states by the living whole of which that matter is a part. The whole forms its own specialized matter. Furthermore, such complex matter does not just come to be under the guidance of the whole; it also *acts* in function of the living whole, in function of its animation. The way the parts of a cell function is the effect of the body acting as a whole. There is a downward causality from soul to the matter that it composes as a living whole.

In other words, in a living organism, the whole regulates the parts and their activities. It is not the case that one part of the organism regulates another—that one neuron does something to another, or that one part of the brain acts upon another. Rather, the whole regulates all the parts. That is what animation means, that a whole regulates itself and all its parts.

To clarify the manner in which the material components of a living thing are affected by the whole, consider the difference between a living being and an artifact, such as an automobile. The materials in an automobile are not changed by the automobile as a whole. The steel remains steel, the plastic remains plastic, the rubber remains rubber, no matter what happens to the car and no matter what the environment is like. But in a living being, the whole entity changes the matter that it takes into itself: it shapes its own bones, muscles, nerves, blood, and enzymes, and it is constantly reconfiguring and activating the matter in its cells. For an automobile to be like a living thing, it would have to change its own matter depending on its surroundings and its age and what it had to do at the moment. It would have to regulate itself and heal itself. Cars do not adapt in that way because they are not self-regulating, living things. Cars do not heal themselves. Much to our inconvenience, a car cannot repair a flat tire the way our bodies heal a wound. Cars are not animated, they do not have soul, but living things do compose their own matter and guide their own development precisely as animated.

And animation also means that the entity acts and interacts outside of itself as a whole, that it does things that the elements could not do by themselves or in combination. An animal attacks its prey, it nurses its young, it builds a nest, it learns to fly, it is fed by its parents. These activities are not merely the resultants of the elementary forces; they are new kinds of activities that are proper to the entity as a whole. The activities of plants, animals, and human beings are specific to the things in question, each taken as an irreducible whole. And in human beings, we have not only the animation of a body but the life of a spirit, a rational and responsible life.

Spiritual Activity in Human Knowing

I would like to convey in an intuitive way what we mean by spirit and its life. We should not take spiritual activities to be something ghostly. It is not the case that spiritual things are given to us only through introspection or through self-consciousness or feelings. To say this would be to speak in a Cartesian way: everything spiritual would be inside. Rather, spiritual activity is present whenever we do things that escape the confinements of space, time, and matter. We do this all the time, and we do it in a public way.

For example, when we rationally communicate with one another, we carry on a spiritual activity, because we share a meaning or a thought or a truth with other people at other places and times. The same meaning, the same thought, the same intellectual identity, can be shared by many people, and it can continue as the same truth over centuries of time, when, for example, it is written down and read and reread at different times in history. Such a truth transcends both space and time, and it transcends material causality as well, because it is the kind of thing that matter alone does not generate. The same truth can be found in many places and in many minds. Also, a cultural object like a drama is a spiritual thing. Shakespeare's *Macbeth* has been achieved over and over again in many places and times when it is performed, read, and interpreted, and yet it is always the same thing. It transcends its embodiments, even though, being something human, it needs its embodiments to be realized. Mathematical formulas, recipes for food, machines, furniture, clothing, flags, political actions, all are spiritual things at least in part. Even something like a ruin of a castle or a crumbling ancient temple has a spiritual aspect, because such things show the presence of reason even while they are being reclaimed by space, time, and matter, and the traces of spirit in them are slowly vanishing. There is something bittersweet about such things, as the signatures of reason in them are gradually extinguished. Human beings saturate the world with spiritual accomplishments, and in doing so they transcend their bodily existence.

Human spirituality can be present in the things that people say and do and make; it is made present in words and machines and furniture, but it is also present in a much more intense way in the human body itself. We have seen that all forms of soul transform the body they animate, and in human beings the rational dimension of soul also exercises downward pressure on the matter that it informs. The human body is made to be an ingredient in a spiritual life. The human brain, for instance, is considered the most complex nonlinear system in the universe,[6] but we should not think that human rationality and spirituality come simply from material causes in the brain. It is more appropriate to say that human rationality somehow shapes the brain; the brain itself is formed by what a human being is and does.[7] And not only the brain, but also the human face and the hand, are shaped by what we are and what we do, as well as by the material we are composed of. Human emotions, desires, and self-awareness are also modified by the fact

---

[6]See Wilson, *Consilience*, 97: "Overall, the human brain is the most complex object known in the universe—known, that is, to itself."

[7]See Calvin and Bickerton, *Lingua ex Machina*, 146: "... the newly emerged syntax would itself have acted as a selective process, tilting the balance in favor of any changes in the nervous system that would lead to the construction of more readily parsable sentences."

that they are elevated by reason. The *telos* of all these things is not just mechanically caused. The human body is enlivened and formed by a soul that permits rational life.[8]

The human body is also affected by the rational interactions that human beings enter into. The human brain and nervous system, for example, can function within human language

---

[8]The contrast between living things and artifacts can be used to shed some light on the dualism that some thinkers introduce between body and soul. Plato, for example, claims that the soul can exist apart from its bodily material, and that it can even migrate from one body to another. Aristotle insists on a close unity between animation and the animated body, but even he speaks of a part of reason (and hence of what we have called spirit) as coming from outside. But St. Thomas Aquinas, in his controversy with the Latin Averroists, says that reason belongs to the individual human soul itself, and stresses even more than Aristotle the unity of the human being. In artifacts, however, there is a dualism between the traces of spirit and the matter that embodies them. Of course, the artist or the maker wants as good a fit as possible between the matter and the form and function he puts into it. The carpenter wants the wood to be appropriate for the table; the artist wants the stone to be the right material for the sculpture. Still, the matter is prepared *before* it enters into the fabricated object; the artistic form does not animate and determine the matter the way the living form shapes its own matter. Even if the fabricator makes a special matter for the artistic object, such as Corian for countertops or concrete for a building, it is the maker, and not the form, that makes the matter what it is. The form remains somewhat alien to the matter, because the form of the artifact is the work of the spirit of the artist, not immediately of his soul. The artist produces the work through his thinking and choices, not through generation.

Consider, furthermore, the difference between the demolition of a fabricated object and the death of a living thing. Suppose a statue is smashed into small pieces, so that nothing of the statue's form remains. The pieces of marble continue to be marble, because they were not constituted as marble by the form of the statue. But when an animal dies, not only does the animal cease to be, but its matter decomposes and ceases to be the kind of matter that it was when the animal was alive. Philosophical dualists interpret living things after the fashion of fabricated things, and take the body as only externally related to its animation. In the case of the human being, they take the soul to be essentially the spirit and see it as present in the body as the artistic form is present in its matter. The matter can go its own way and leave the spirit or soul unaffected. The problem of dualism was introduced at this conference during the discussion period by a question from Francis Cardinal George, O.M.I., who also raised the issue discussed below in note 14.

only if such language is learned during a specific period, a window of opportunity in the child's development. If the child is deprived of human linguistic interaction during that critical age, the person will not become able to learn to use human grammar. He will be stuck at the level of what is called protolanguage, not language itself.[9]

In Thomistic philosophy, the rationality of the human person is usually associated with the ability to think about all forms of being and to achieve universal concepts.[10] The universality of human understanding is the central phenomenon used to show the rational and hence spiritual nature of the human soul. Such universal thinking is able to escape the confinements of space, time, and material causality. In it we adopt a standpoint that is not limited in the way sensory experience is limited. This is a good approach to human rationality, but other phenomena can also be introduced to confirm it, such as our ability to use language with its grammar, the ability to intend things in their absence, and phenomena like picturing and quotation, as well as the specifically human form of remembering the past and anticipating the future. All these topics can be developed to show in greater detail the transcendence of the human person.

Spiritual Activity in Human Willing

I have discussed briefly the way human spirituality expresses itself in cognition and communication. Let us consider the way it appears in volition, in human responsibility and freedom. In normal circumstances, people are held responsible for what they have become, and such accountability is not found in the case of animals or plants. A particular deer, for example,

---

[9]On the difference between language and protolanguage, see Derek Bickerton, *Language and Species* (Chicago: University of Chicago Press, 1990). Bickerton claims that protolanguage is an entirely different kind of communication than language; the latter includes syntax and the former does not. Protolanguage occurs in children under two years of age, and people who fail to develop real language may be lodged in protolanguage for their entire lives. See 110–118.

[10]See, for example, Kenneth Schmitz, "Purity of Soul and Immortality," *Monist* 69 (1986): 396–415. For a comprehensive study of Thomas's doctrine on the immateriality of the human intellect, see Michael J. Sweeney, *Thomas Aquinas's Commentary on De Anima 429a10–429b5 and the Argument for the Immateriality of the Intellect* (Ph.D. diss., The Catholic University of America, 1994.) This work includes an extensive bibliography.

may be the best in the herd or it may be one of the inferior members, but in neither case do we say the deer is morally responsible for being what it is. The deer may have had to fight to assert its status, but the question of moral responsibility does not arise. The animal does not *have* its nature the way a human person has his. But if a mature human being is courageous or cowardly, temperate or intemperate, energetic or lazy, generous or avaricious, we acknowledge that he is that way through choices he has made. He has shaped his own life, and he has done so in a deeper and more personal, more spiritual way than even his soul has shaped him. He is what he is not just because he was born a human being, nor just because he lived in this particular situation, but most of all because he did what he deliberately chose to do in his circumstances. He lived his life, and continues to live it, with responsibility, because he can know what he is doing and can make choices to determine not only the world around him but himself as well. We define ourselves within our humanity because we are rational, because we are spiritual. We may not be famous, we may never be mentioned in the newspapers or on television; but each of us lives his own life and answers for it, and in doing so he acts spiritually. Animals and plants do not do this, and hence they lack the dignity as well as the responsibility of persons.

This spiritual responsibility of persons shows up vividly for us in their benevolence and malevolence, in the good or bad things they do to us. If an animal attacks us, we consider ourselves unfortunate, but we do not bear resentment toward the animal; we do not think that we have been treated unjustly or malevolently. But if a human being deliberately injures us, we do bear resentment because we see that this agent did what he did through knowledge and choice.[11] He understood us as someone to be harmed, and it was through that understanding that the injury was done. It was a spiritual and not just a natural infliction of harm. Human benevolence and generosity, and human friendship, are also expressions of spiritual rationality.

Along these lines, human political life is also brought about by man's spiritual nature. People form political societies not just because they herd together like animals, but because they

---

[11]See Robert Sokolowski, *Moral Action: A Phenomenological Study* (Bloomington: Indiana University Press, 1985), 56–57. Also, "What Is Moral Action?" in *Pictures, Quotations, and Distinctions: Fourteen Essays in Phenomenology* (Notre Dame: University of Notre Dame Press, 1992), 261–276.

share an understanding of the good, the noble, and the just; and this common understanding is the basis for a life in which people can pursue goods that are common to the entire community, not just individual goods. Consider how complex political life is, and how intricate the decisions that are made in it. Such decisions are public achievements and they are done by many people working together. They could never be reduced to the materiality of the brain or the body alone. As John McCarthy writes, "In what way ... could the known laws of physics even begin to explain Abraham Lincoln's decision on March 12, 1864, to appoint General Ulysses S. Grant commander of the Union armies? What would be the shape of the formula astute enough to recognize [General George] McClellan's performance in that office as a failure, to say nothing of a formula powerful enough to predict that failure?"[12] How could Lincoln's decision ever be explained as the resultant of a complex network of purely material forces? The action is public, and it can only be described and explained in terms of rational, spiritual categories, not material ones. The action can only be explained as the performance of a whole individual with spiritual powers, the performance of a human person.

It is an extremely curious thing that human beings spend so much energy denying their own spiritual and rational nature. No other being tries with such effort to deny that it is what it is. No dog or horse would every try to say that it is not a dog or horse, but only a mixture of matter, force, and accident. Man's attempt to deny his own spirituality is itself a spiritual act, one that transcends space, time, and the limitations of matter. The motivations behind this self-denial are mystifying indeed.

Spirituality in Religion and Christian Faith

The fullest way in which human spirituality is visible is found in the practice of religion. Religious ceremonies and activities are specific to human beings; animals do not have processions, rituals, or hymns. Religion is the apex of human rationality, because it deals with the highest truth and the greatest good. It is man's response to what is first and best in the world or beyond the world, and it sheds light on all the other

---

[12]John C. McCarthy, "The Descent of Science," *Review of Metaphysics* 52 (June 1999): 846. This essay is a review of Wilson's *Consilience.*

things that man knows and loves. If the exercise of truth is a spiritual activity, then our response to the source of truth is even more spiritual. This religious striving toward the first and the best is found in all human beings, even those who are materialistic reductionists. They too find in evolution a kind of supreme, quasi-providential power that guides the development of the world. We have already mentioned Derek Bickerton's claim that evolution made us what we are and, for good reasons of its own, provided us with language. In a recent report about the role of sleep, the neuroscientist Terrence J. Sejnowski says, "Why do almost all of us need eight hours of downtime each night? Our sensory systems are down, our muscles are paralyzed and we are very vulnerable. Evolution must have had a purpose in mind."[13]

The various religions in human history are different ways of articulating the whole of things and of responding to the divine power or powers that govern the whole. As Christians, we believe our faith is more than one religion among many. We believe that in the events of the Old Testament God intervened providentially in the life and events of the Jewish people, and that in Christ he not only spoke to us but became one of us, and thereby elevated human reason and human spirituality to a level beyond what it could reach by its own efforts. Reason and spirit become deepened in the words and actions of Christ, who told us of a truth that exceeds anything we could naturally attain, revealed a divine love that is stronger than any human friendship, and disclosed an eternal life that is more than we could hope for in our natural condition. Human spirituality, the elevation of the human body into a life of reason and spirit, is so much deepened by the work of Christ that the human body, even as a physical thing, can now share in the resurrection and eternal life. The fact that the human body can enter into the spiritual activities of rational thinking and responsible action provides a kind of apologetics for the possibility of the resurrection of the body, in which, through the saving power of God, the body is even more intensely elevated into a spiritual condition.

---

[13]Sandra Blakeslee, "Experts Explore Deep Sleep and the Making of Memories," *New York Times*, November 14, 2000, F2.

Witness

I wish to close with a practical recommendation. Our modern culture tends to think of human beings as merely material things, and it encourages them to act accordingly. The Church, through her teaching, liturgy, and practice, is a witness to the spiritual nature of man, to his personhood. She is a witness not only to the Word of God but also to the rational nature of man. It is the mission of the Catholic educational system, on all levels, to be involved in this witness, to help students to become more clearly and reflectively aware of human spirituality, to teach them how to think about human beings in a way that respects their personhood, to provide them with categories appropriate to human nature in all its fullness, and to convey to them the true sense of teleology that must become part of our understanding of the world.[14] T. S. Eliot formulates this goal in these words:

> The purpose of a Christian education would not be merely to make men and women pious Christians: a system which aimed too rigidly at this end alone would become only obscurantist. *A Christian education would primarily train people to be able to think in Christian categories*, though it could not

---

[14]Some valuable work on the concept of evolution has been done by Richard F. Hassing. See "Darwinian Natural Right?" *Interpretation* 27.2 (Winter 2000): 129–160, and "Reply to Arnhart," *Interpretation* 28.1 (Fall 2000): 35–43. Hassing asks: (1) Must we say that there is one good that all living things seek? Is the good for all living things transspecific, the same for all species? (2) Or must we say that living things develop in such a way that they seek goods that are species-specific, that is, goods that are proper to the species in question, including the human species? In the first option, all apparently "higher" goods would be explainable as strategies for survival, which is the one good common to all living things and the one guide for all evolutionary development. In the second option, the possibility arises that there are goods specific to human beings, such as truth, moral goodness, and nobility. Hassing has also developed the important concept of teleology, or the natural ends of things, in a work that he edited under the title *Final Causality in Nature and Human Affairs* (Washington, D.C.: The Catholic University of America Press, 1997). I would call attention especially to the contributions by Hassing and Francis Slade. See also, Francis Slade, "On the Ontological Priority of Ends and Its Relevance to the Narrative Arts," in Alice Ramos, ed., *Beauty, Art, and the Polis* (Washington, D.C.: American Maritain Association, 2000; distributed by The Catholic University of America Press), 58–69.

compel belief and would not impose the necessity for insincere profession of belief.[15]

To help people be able to think in Christian categories is a difficult task, since the categories that our culture provides are very different, especially the ones it offers us for thinking about human beings. It is important for Christian schools and universities not only to teach but also to develop, through research and writing, a deeper understanding of human spirituality, and to formulate it in terms that are appropriate for our day and age, in which we take into account evolution and neuroscience and molecular biology, as well as artificial intelligence and computer science. I think that many, if not most, Catholic colleges and universities have lost the distinctiveness of their mission in the liberal arts and sciences. A major reason for this loss is the fact that in the past forty years Thomism ceased to be the unifying focal point, and nothing has taken its place. A revival of what I would like to call "streamlined Thomism" would be very appropriate in Catholic education today, along with a special focus on the human soul, the human spirit, and the human person. Such an effort would be in keeping with the vision expressed by the Holy Father in his encyclicals *Fides et ratio* and *Veritatis splendor*, as well as his Apostolic Constitution *Ex corde ecclesiae*. The effort is called for not only by the needs of the Church, but by the perils to human nature and the human person that are arising in our contemporary world. It would be of great strategic importance in dealing with questions of medical technology, bioethics, the biological and spiritual origins of the human person, and the dignity of human life, from conception to natural death. The Church could perform a great service to all people of good will if she were to offer them a deeper and more spiritual way of thinking about the human person.

---

[15]T. S. Eliot, "The Idea of a Christian Society," in *Christianity and Culture* (New York: Harcourt, Brace and World, 1940), 22. Italics added.

HISTORICAL PERSPECTIVES ON THE HUMAN PERSON

ELIZABETH FOX-GENOVESE

What is man that thou art mindful of him? And the son of man that thou visitest him?

For thou has made him a little lower than the angels, and thou has crowned him with glory and honor.

Psalm 8:5–6

Contemporary Western society—the most materially advanced countries in the history of the world—stands alone and without precedent in the high value it attributes to the individual person. Simultaneously, it stands exposed for the cheapness in which it holds human life. Individual rights, human rights, self-esteem, and related concepts dominate our culture's sense of the good that must at all costs be defended. Yet unborn babies, terminally ill patients, or those who simply "dis" others in the street are deemed expendable. Some lives embody the essence of all that is admirable and worthy; others are to be brushed aside as mere encumbrances. What remains to be explained is who gets to decide which lives deserve respect and protection and which do not? Which of us has a right to decide which lives are worth living?

The well-known passage from the eighth Psalm with which I began reminds us of the unique place the human person enjoys in creation, delicately poised between God, whom we are made to serve, and other living creatures—animals, fish, and birds—over whom God has granted us dominion. Contemporary culture, certainly in the United States and Europe, readily embraces the idea of man's dominion, but it shows markedly less enthusiasm for the idea that we rank lower in the hierarchy of merit than the angels and God. Our age has lost the Psalmist's marvel at the unique blessings that God has showered upon us, preferring to assume that they are ours by right or by our own merit. Our complacency and self-satisfaction constitute the very essence of the culture of death against which the Holy Father warns us, for our boundless self-absorption blinds us to the value of others.

Caught in a dangerous paradox, our age simultaneously celebrates the unique value of human life and, however inadvertently, dismisses it as of no consequence. The life we are told to value is our own, and the more highly we value it the more easily we are tempted to discount the value of the lives of others. Preoccupation with self at the expense of the other is nothing new: Cain established the model at the dawn of time. But our culture is breaking new ground in its attempt to establish selfishness as a higher principle, swathed in words like choice and fulfillment and autonomy.

In historical perspective, fixation upon the rights and unique value of the individual is something new. Until very recently, societies, including the most sophisticated societies of the Western world, have primarily regarded individual persons as members—or articulations—of groups, notably as members of families, but also of clans, tribes, social castes or estates, religious orders, or various trades or professions. The preferred forms of classification have varied, but the prevailing principle has held firm. A human person has been understood as someone's daughter, father, wife, or cousin—one link in a kinship that defines all of its members.

Christianity's insistence that God loves each individual broke radically with these patterns. Christianity affirmed the value of each particular person independent of ethnicity or sex or social standing, pronouncing, in the words of St. Paul, that "There is neither Jew nor Greek, there is neither bond nor free, there is neither male nor female: for ye are all one in Christ Jesus" (Gal 3:28). More, in affirming the value of each, Christianity also affirmed the value of all. In other words, Christianity viewed the human person as both particular and embodied and as universal. The parable of the Good Samaritan was intended to teach Jesus' followers that the command to treat others with charity extended beyond the members of one's own ethnic group.

In the Christian perspective, it was not possible truly to value any single person without valuing all persons or to value all persons (humanity) without valuing each single person. In both respects, Christianity broke with the tribalism of ancient Israel and of much of the rest of the world, establishing new standards for the freedom each person should enjoy and for equality among persons. Christians did not, however, immediately attempt to impose their standards of spiritual dignity and spiritual equality upon relations among persons in the world. Over time, Christianity powerfully influenced the character of Western culture and even political life, but it owed much of its success to its remarkable ability to adapt to prevailing institutions and relations.

Only with the birth of modernism, notably in the notorious—if widely celebrated—*cogito ergo sum* of René Descartes, did the disembedding of the individual person from the collectivity that grounded his identity begin to be viewed as a positive good. And only with the Enlightenment and the great eighteenth-century revolutions, notably the French Revolution of 1789, did the ideal of individual freedom attain preeminence over all forms of dependence and connection. In the waning years of the eighteenth century a political and intellectual vanguard proclaimed freedom the absolute antithesis of slavery and promulgated an understanding of freedom that favored the severing of all binding ties among human beings. Freedom in this lexicon means autonomy, self-determination, and independence from binding obligations unless freely chosen, and this is the idea of freedom that has triumphed in our own time. Significantly, it originated as a radically secular idea, one frequently launched as a direct challenge to God. At the extreme, its consequences have been disastrous, but its most chilling implications may yet lie ahead. For it is the pursuit of this ideal of freedom that has brought us sexual liberation, abortion, assisted suicide, and an entire battery of assaults upon human life.

Before focusing upon the ways in which the radical pursuit of freedom has cheapened the value of the human person, however, it is necessary to acknowledge its many benefits. For the pursuit of human freedom has heightened the dignity and improved the lives of countless persons throughout the globe.

The same history that has brought us the progressive discrediting of ties among persons has promoted a remarkable improvement in our understanding of the intrinsic value and dignity of each person. During recent centuries in many parts of the world, we have witnessed the abolition of slavery, an improvement in the position of women, greater attention to the discrete needs of children, respect for the needs and dignity of those who suffer from various handicaps, and more. As Pope John Paul II has emphasized, these gains are not trivial, and on no account must we countenance their reversal. The puzzle remains that they have been accompanied by—and many would argue have depended upon—a hardening of attitudes towards the intrinsic value of all human persons and, especially, towards the binding ties among persons.

These two tendencies confront us with a paradox. On the one hand, we have a decreasing respect for the bonds among persons, and on the other an increasing commitment to the value and rights of previously oppressed groups of persons. On the one hand we have an inflated concern for the rights and sensibilities of the individual, on the other a callous disregard for any life that, in any way, inconveniences us. This paradox challenges us to rethink our understanding of the human person and, especially, the nature of the freedom and rights to which each of us is entitled. For a misguided understanding of freedom will inescapably shape our understanding of the claims of life. Presumably, if one values human life, one opposes its willful termination, especially in its most vulnerable forms. Yet many of those who claim to value human life view abortion, assisted suicide, and even infanticide as necessary to its defense. For, only the right to secure liberation from unwanted obligations protects the individual's freedom, which many view as the essence of any life worth living.

There is nothing surprising in the inclination to view freedom as freedom from rather than freedom for, with the "from" understood as oppression and the "for" understood as service. Throughout history, the majority of labor has been unfree and the majority of women have been subordinated to men—initially their fathers or uncles and, later in life, their husbands or brothers. The Old Testament and classical literature both abound with examples. Agamemnon sacrificed his daughter, Iphigenia, to further his prospects for victory in the war against Troy. Until recent times, Hindus in India practiced suttee whereby a widow was burned upon her husband's funeral pyre. Even in England, the sale of wives, although increasingly rare, persisted into the nineteenth century. Similarly, serfdom and slavery persisted throughout the world well into the nineteenth century and may still be found in some places today.

Under conditions in which even upper-class women rarely owned property in their own names and poor women might be beaten or bullied at will, it is not surprising that the early proponents of women's rights embraced the analogy of slavery to describe their own condition and spoke of breaking the chains of their bondage. In practice, the women who were most likely to protest women's condition were from the urban middle class and sought to enjoy the same advantages of education and professional careers as their brothers. Such women normally benefited from codes of middle class gentility and did not suffer from the horrors of abuse, sexual slavery, oppression, denigration, and desertion that plagued less privileged women—although some did. But they readily depicted their lot as indistinguishable from that of their less fortunate sisters. By the early twentieth century, the more radical were beginning to argue that marriage and childbearing were the true seedbeds of women's oppression and to lobby for expanded legalization of divorce and artificial contraception.

Throughout these and related efforts, feminists continued to describe their goal as freedom from bondage and to claim their right to be regarded and treated as full human persons. Most people initially responded to the women's movement with hostility or incredulity. Even among opponents of the women's movement, few claimed that women did not count as full human persons, although most insisted that they were very different persons than men and, consequently, in need of a different social situation. The conviction that women and men, although both fully human persons, differed by nature persisted well into the twentieth century. Doctors argued that women's bodies made them unfit for college, psychologists argued that women had a distinct criminal disposition, lawyers argued that women should not be admitted to the bar, and countless people argued that women should not vote. Yet within the comparatively brief span of a century or so, feminists began to convince growing numbers of people of the justice of their cause, and, in so doing, to convince many that the natural differences between women and men had been vastly exaggerated.

We would be rash to minimize the magnitude of their extraordinary rhetorical victory, which significantly expanded the meaning of individual freedom and ultimately resulted in a reconfiguration of the moral landscape. By rhetorically extending the absolute opposition between freedom and slavery to the condition of women, feminists had declared any limitation upon a woman's freedom—including those imposed by her own body— as illegitimate. The campaigns against slavery and the subordination of women both embodied a worthy—indeed necessary— commitment to increasing the dignity accorded to all human persons and the equality among them. Both, in other words, represented what we may reasonably view as moral progress. Yet both tacitly embraced the flawed premise that to be authentic, freedom must be unlimited, or better, unconditional, which, by a sleight of hand, reduced the ties among persons to another form of bondage.

Throughout the modern period, material change has undergirded and, arguably, accelerated changing attitudes towards the human person. Urban societies provide many more possibilities than rural societies for people to live alone. In rural society the interdependence of persons constitutes the very fabric of life, and none can survive without mutual cooperation. Typically, rural societies also benefit from a clear articulation of authority whereby one member of the family or group assumes primary responsibility for and direction of the rest. Typically, that person has been the male head of a household, family, or tribe. What modern critics are loath to understand is that rarely—if ever—could such a head exercise his authority without the tacit or active collaboration of those over whom he presides.

The picture of traditional societies is all the more difficult to paint because they are so easy to romanticize. These were worlds in which life for many could often be, as Thomas Hobbes said, "nasty, brutish, and short." They were worlds in which cruelty among persons abounded and in which death stalked young and old alike. It is not insignificant that the Palestine of Jesus' time abounded with cripples and lepers, hemorrhaging women, and desperately ill children. Population has increased in the modern world because modern medicine has done so much to control disease and defer death, rather than because of an increase in the number of births. The prevalence of disease and the likelihood of early death have led many traditional rural societies to value highly women's fertility and the birth of children. Here too, however, romanticization misleads, since even groups that welcome births might turn around and kill infants they could not support. Traditional societies, even when Christian, did not necessarily manifest "respect for life" in the sense we use the phrase today. They did, however, know that their agricultural and military survival depended upon sustaining their population or increasing it.

These traditional rural worlds did not ordinarily celebrate the unique attributes of each person as we are wont to do today, but they did value each person as an essential member of the family, household, or community. No family could function for long without a mother or appropriate female substitute, typically a maiden aunt, and widowers with small children were normally quick to remarry. Similarly, it could not function for long without a strong senior male who, with or without assistance, could bring in crops, care for livestock, and defend against predators. Personal autonomy did not figure as the *summum bonum* among rural folk for whom interdependence provided the best guarantee of family or community survival. Modern critics of those bonds frequently focus upon the injustice of specific forms of subordination: slave to master, wife to husband, children to parents. But in repudiating the injustice of women's subordination to men, for example, they end by demanding women's "liberation" from all binding ties, presumably on the grounds that such ties have historically put them at a disadvantage.

The acceleration of economic progress transformed the rural world, mainly by moving the economic center of gravity to cities, which offered new possibilities for people to live in smaller groups or even on their own. Drawing ever larger numbers of people into wage labor, capitalism insinuated itself into the interstices of families and households, reinforcing the culture's growing tendency to encourage members to see themselves as individuals. Capitalism's dependence upon an accelerating consumption of material goods furthered the cultural emphasis upon the psychological goods of autonomy, satisfaction of desire, and instant gratification. Secular psychology contributed mightily to the transformation of "I want" from evidence of selfishness or greed to a sign of mental health. In the same spirit, it declared war on the idea of sin, which it denounced as nothing more than a sadistic campaign to thwart people's enjoyment of life.

Some modern scholars have delighted in exposing the ways in which traditional societies held persons in thrall to repressive norms, primarily fostered by punitive, misogynist, patriarchal, and hierarchical forms of Christianity—especially Catholicism. Others have tended to romanticize folk and working class cultures, presenting them as more spontaneous and less repressed than the middle class culture of the modern urban world.

Both views contain a measure of truth and a measure of false-hood. But both, however inadvertently, suggest a greater emphasis upon the bonds among persons than commonly prevails today. Whether one views the traditional world as good or bad—or, more plausibly, a mixture of both—it was a world in which people developed a sense of themselves as persons as a function of their relations with others, often beginning with their relation to God. Nor were these attitudes unique to Christians. In different versions, they prevailed among Jews, Muslims, Hindus, and Confucians, as well as among adherents to various forms of polytheism, animism, and other systems of belief.

Here, I do not wish to engage in debates about the deeper character of the various faiths but simply to underscore that all have shared a sense of the human person as an integral part of a larger group whose needs would take precedence over the specific person's whims and desires. Significantly, religion has often exercised its strongest influence among the members of predominantly oral rather than predominantly literate cultures. Oral cultures tend to promote unambiguous messages about good and bad, about heroes and villains. They further attribute little importance to individual subjectivity, and they have little interest in progress. Like all others, traditional oral cultures do change, but since they lack written records, they do not register change but rather absorb it into "the way things have always been."

Traditional societies' conservatism with respect to the rights and independence of the individual, like their strong commitment to holding persons to prescribed social and familial roles, represented above all a commitment to the survival and internal coherence of the family or community. In this spirit, they rejected individual judgment as an appropriate guide for behavior, not least because they fully recognized the disruptive power of individual passions, notably anger and desire. And although to enforce their convictions, they might use methods that seem repressive or even brutal today, they did not often rely upon the extreme measures that some contemporary groups have been known to use such as killing girls who try to attend school. Today's extremes, as Bernard Lewis, the great authority on Islam, has argued, primarily embody a panicky reaction against what are perceived as the excesses of the disintegration and decadence of contemporary Western society.

Those who fear the destructive potential of Western cultures do not err. It is child's play to muster examples of rampant consumerism and what Karl Marx called the fetishism of commodities and what today seems to be degenerating into the commodification of personal relations. Even the most casual acquaintance with American media—especially television—reveals a world that has all but dehumanized persons by severing the binding connections among them. For all the talk of the warmth and support of substitute, alternate, or proxy families, American culture depicts a world in which family represents no more than a person's current choice, which may easily be replaced by another choice. We have never lacked for critics of the symptoms of this culture of easy-come, easy-go; and recent years may even have seen an increase in the defense of marriage, attacks upon the harm that divorce wreaks upon children, and defense of the value of modesty and premarital chastity. We have dedicated groups and individuals who oppose abortion and so-called assisted suicide, and we are apparently seeing an appreciable increase in the number of Americans who have doubts about the desirability of legal abortion, especially after the first trimester. What we lack—and the lack is devastating—is an insistent, concerted counteroffensive. We lack it because more often than not even those who oppose many specific forms of social decomposition accept the main premise that underlies them all, namely the primacy of the convenience and comfort of the individual.

Recently, I had the opportunity to speak with a group of faithful Catholic women, many of whom attend daily Mass, all of whom rightly consider themselves devout. Blessed with considerable material comfort, they have all given generously to the parish for decades, and would all consider themselves loyal supporters of the Church. During my talk, I mentioned my growing horror at abortion as one of the important elements in my own conversion to Catholicism. When we broke into informal conversation over lunch, one of the women drew me aside because she wanted me to know that notwithstanding her respect for the teachings of the magisterium, if she had a thirteen-year old daughter who was impregnated during a rape, she would whisk her off to an abortionist before you could say "boo." Startled, I responded, "And what if she were twenty-three and finishing law school?" She looked startled and suddenly abashed.

An intelligent woman, my acquaintance readily understood my point that, with each passing day, we Catholics seem to be finding it easier to acquiesce in the logic of the secular world. The results are disastrous for our understanding of the human person and our ability to sustain binding relations with others. We have too readily acquiesced in the secular view of the human person as, above all, an individual who is fundamentally disconnected from other individuals. The disconnected individual may enter into relations with others, but the relations remain contractual, subject to termination at the choice of either party. Such instrumental relations are bad enough for adults, but they are disastrous for children, not to mention the handicapped, the unborn, and the terminally ill. More, they are in direct contradiction to the teachings and spirit of our faith. For how are we to understand Jesus' repeated commands that we love one another as ourselves if not as a command to recognize that our own personhood depends upon our recognition of the other. In this sense, the modern era has not merely transformed the idea of the person, it has effectively abandoned it in favor of the subjective individual.

The irony of our situation is painful. In historical perspective, we appear to place a higher value on the individual—as an individual—than any previous society, yet we increasingly view the individual as essentially a subjective being whose will and desires should determine what he or she is due. Only in this spirit would it be so easy to present an unborn baby—and, for some people, one that has been born as well—as nothing more than an obstacle to its mother's freedom to pursue the goals she has chosen. Rather than emphasizing the mother's obligation to the human life she is carrying, our culture increasingly insists upon her right to be free of it. The mother's right to choose thus negates the unborn child's and by claiming this right to deny the personhood of another, the mother negates her own. No longer a person whose being, sense of self, and place in the world depend upon her relations with others, beginning with her primary relation with God, the woman becomes an isolated individual, disconnected from others whom she can see only as objects to be manipulated or obstacles to be cleared away.

Christianity led the way in promoting a view of each person as valuable, unique, beloved, and endowed with freedom. Yet Christianity also presupposed that each person derived meaning from relations with others—that the very essence of personhood lay in the recognition of the equal personhood of the other. Thus, the Christian ideals of the value and freedom of each individual coexisted comfortably with a culture that placed a much higher value on the group than the individual, who was primarily understood as a member of the group. The Reformation placed a new emphasis on the individual, but Luther, Calvin, and their heirs remained tied to a communal ethos. Their doctrine of *sola scriptura*, however, opened the way for a disastrous slide into a rising secular bourgeois individualism that, in our time, has largely overwhelmed the Protestant churches and—let us admit frankly—is now threatening our own. For as individualism gradually triumphed over the collective values of traditional culture, it did so in radically secular terms and usually in direct rebellion against the Church.

This history has left us a dangerous and insidious legacy, for the individualism that spearheaded a broad cultural revolt against the teachings of the Church also insinuated itself into the thinking of Christians, including Catholics. The goods that individualism purported to offer are almost irresistibly seductive: tolerance of the behavior of others; delight in bodies and sexuality; acceptance of oneself, complete with flaws; the legitimacy of desire; and on and on. Consequently, any attempt to oppose or criticize them seems ungenerous, judgmental, and intolerant. Who am I to tell another how to live his or her life? What gives me the right to impose punitive values upon another?

This purported and seductive solicitude for the freedom of each individual masks the ominous tendency in our Western societies to objectify the very individuals we pretend to celebrate. Substituting rights for mutual bonds, we are substituting the individual for the human person, thereby freeing ourselves to deny the humanity of others. Thus does the slaughter of the innocents become "a woman's right to choose."

Genetics
and Embryology

GENOMICS AND ETHICS: WHERE THE MAIN ACTION IS TODAY

LUKE GORMALLY

Practical Implications of the HGP and Some Ethical Issues They Raise

Genomics is the study of the human genome. It embraces both theoretical knowledge of the genome and what we can do with that knowledge. There are many claims about the practical implications of that knowledge. Here are a few examples:

- It is expected that it will be possible to use genetic analysis in clinical practice to predict those who are at risk of various diseases, just as blood pressure measurements are used nowadays to predict risk of cardiovascular disease. Having identified those who are at risk, one may either adopt preventive strategies or, at least, be better placed to detect the early onset of disease and employ the best approach to its management.

- It is claimed that a number of conditions, in particular single gene disorders, can be treated by somatic gene therapy. Clinical scientists have been making optimistic claims about the advent of such therapies for some time. When the cystic fibrosis gene was identified in 1989, we were told that gene therapy was around the corner. To the best of my knowledge, we still have no treatment for the condition based on gene function. (I notice that Dr. Coors mentions in her paper in this volume the recent announcement by researchers of "credible successes of gene therapy in improving the health of Hemophilia B patients and Human Severe Combined Immuno-deficiency-X1 Disease," but it is not clear that these procedures are proven to be safe.)

- It has been proposed that we can intervene to prevent the continued transmission of certain genetic disorders from one generation to the next by germ line gene therapy, that is, by the genetic modification of sperm or ovum, either directly or by gene transfer to the early embryo.

- Some anticipate modifications of the human genome designed not to correct malfunctioning or eliminate disease but rather to improve or enhance abilities which already fall within a normal range. People have in mind a range of abilities, from physical abilities associated with sporting achievement to intellectual abilities associated with academic achievement.

- It is widely recognized that the greatly increased knowledge of the human genome has considerably increased the ability to detect abnormalities in utero with a consequent increased likelihood of people choosing to end the lives of children with such abnormalities by abortion.

- In so far as genomics results in people obtaining comprehensive genetic profiles on themselves, it will both increase the burden of choice about having children, as well as making possible the avoidance of conceptions which are likely to result in children bearing a heavy burden of disease. Presumably genetic counselling will be much more in demand, either by couples deciding whether there are genetic reasons for them not to marry (as, for example, when they are both carriers of the abnormal autosomal recessive gene for cystic fibrosis) or genetic reasons for them not to conceive when they are already married.

- It is already the case that the pharmaceutical industry is using advances in genetics to aid drug discovery. It is anticipated that drugs will be created based on the proteins, enzymes, and RNA molecules associated with genes and diseases.

- The new knowledge of genetics is not merely aiding the discovery of drugs but is expected to permit the more accurate design of therapies by improving knowledge, both of the movement of drugs through the body from absorption to elimination (pharmacokinetics) and of drug toxicity. Among other claimed benefits, this is expected to permit the marketing of previously sidelined drugs for those who can be predicted not to have adverse reactions to them.

- The pharmaceutical industry is highly competitive. One way individual companies have sought to secure advantage over rivals is by patenting genes, that is, sequences of DNA. Until recently, patent law assumed a distinction between, on the one hand, discovering a part of nature and, on the other, inventing something new. An invention could be patented, a discovery could not. Lawyers and legislators have connived in the subversion of this distinction. The effect of that connivance is that there are now over four million patents on human genome sequences. The company Human Genome Sciences has patented the entire genome of a number of bacteria, including one that causes meningitis. The legal effect of such moves is that others cannot work with the patented genes without the permission (which, if it comes, may come at a price) of those who have the patents. But the sequencing of a gene might have a number of applications, such as the development of diagnostic tests or of gene therapies or of pharmaceutical products. Those who possess the patent may be interested in only one application while being in a position to block other applications.

- The genetic profiling of individuals should allow doctors to prescribe the best available drug therapies for individuals.

- It is expected that genomics will have an enormous impact on the insurance industry, and therefore on insurance coverage for healthcare. For one thing, if insurance companies are able to obtain genetic profiles of populations, they will gain a much more accurate picture of the likelihood of diseases and of their likely liabilities. For another, if they obtain genetic profiles of individuals, then those thought to be at significant risk of suffering from certain conditions may either be refused insurance coverage or may be charged very high premiums.

The above listing of some practical implications of genomics—a far from comprehensive listing—suggests a host of associated ethical issues:

- Questions concerning distributive justice arise in connection with the control of scientific knowledge and the development of therapies by the pharmaceutical and biotechnology companies.

- Issues related to the right to privacy arise in relation to proposals to disseminate information about an individual's genome.

- Issues arise concerning possible unjust discrimination against individuals by employers and insurance companies with access to genetic information about those individuals.

- There are issues about the burdens and benefits of somatic gene therapy and in particular about the dangers of exposing patients to disproportionate risks in the development and application of gene therapy.

- There is the issue of whether one has the right to seek to modify the germ line, particularly when the consequences for subsequent generations are unpredictable.

- There is the question about whether there can be any justification in principle for proposals for genetic enhancement which are unconnected with therapeutic benefit.

- There are questions about how directive and how non-directive genetic counselling should be.

- There is the issue of abortion associated with genetic screening.

- There are questions about the justifiability of invasive techniques for genetic screening of the embryo even when there is no intention to abort.

- There are questions concerning the acceptability of

employing the reproductive technologies to obviate the transmission of genetically-linked conditions.

- And there is the issue of the destructive use of human embryos in genetic research.

Where the Main Action Is Today

Some of the principal ethical issues raised by genomics are to be given in-depth discussion in the other chapters of this book. So it is important that I do not trespass on the work of those authors. It was suggested that I should take a "broader sweep" over the topic area. High altitude aerial reconnaissance has its uses, but the ethical landscape in the field of genetics is so varied that a panoramic survey in a chapter as short as this could only be breathtakingly shallow. So it seemed more profitable to come down to earth in one part of the territory and cover some issues which are not obviously covered in any other part of this volume.

There is reason to think that for the foreseeable future the biggest social impact of genomics will not be in the development of therapies but in the vast expansion of screening programs aimed in one way or another at preventing the birth of children with genetic defects. A friend, who is director of The Advanced Biotechnology Centre at one of the major university medical centers in London, observed recently that, on the evidence of international conferences he attends, this is where the main action is for the immediate and medium-term future.

Now it ought to be obvious that the Catholic community, at least in North America and Western Europe, is not well placed to meet this challenge—for reasons to be remarked on later.

An increasing number of ordinary people are under pressure to make choices which will decisively shape the character of their own lives and the lives of their children: choices about the transmission of human life in the light of increasing knowledge of the transmission of genetically-based abnormalities. There are a number of kinds of such choice which are now proposed as clinical possibilities:

- choices to abort an already conceived child in the light of a prenatal diagnosis of some abnormality;

- choices, in anticipation of a quantifiable risk of transmitting an abnormality, to have one's offspring generated in vitro with subsequent discarding of any embryos shown by embryo biopsy and genetic tests to be abnormal, allowing only a normal embryo or embryos to be placed in the womb;

- choices, again in anticipation of quantifiable risks of transmitting abnormalities, to use, in in vitro fertilization, germ cells which are only partly one's own. For example: in some genetic diseases the genes at fault are in the mitochondria of the ovum, that is, the DNA-bearing structures outside the nucleus of the ovum. One suggested approach to obviating the transmission of such diseases is to take the nucleus from the ovum of the woman who is the carrier of the disease and transplant it into a donor ovum with healthy mitochondria after the removal of the nucleus of the donor ovum; and

- there are many other so-called "reproductive choices" proposed to people who, in the normal way of conceiving, might transmit a genetic disorder: choices, for example, to use donated sperm or ova.

It is hardly possible to overestimate the cultural and societal pressures brought to bear on potential parents to have normal children. Professor Robert Edwards, the embryologist who, together with Patrick Steptoe, pioneered in vitro fertilization, was quoted in *The Sunday Times* (of London) as saying: "Soon it will be a sin for parents to have a child that carries the heavy burden of genetic disease. We are entering a world where we have to consider the quality of our children."[1] Genetics has always been associated with eugenics, the movement to improve the quality of the population by discouraging the procreation of children by those deemed to be unfit and by encouraging procreation by those deemed to be of superior quality. Francis Galton who coined the term "eugenics" (he was a cousin of Charles Darwin), recommended, in his book *Hereditary Genius* (1869), judicious marriages between men of genius and women of wealth! The favored tools of eugenics in the early part of the twentieth century were immigration control and forced sterilization. By 1931 most States of the Union had sterilization laws aimed at eliminating the mentally retarded, epileptics, and sexual deviants. Members of German medical faculties, many of which had institutes of race hygiene well before the Nazis came to power in 1933, would deplore the extent to which Ger-

---

[1]L. Rogers, "Having Disabled Babies Will Be 'Sin' Says Scientist," *Sunday Times* (London), July 4, 1999, News 28.

many lagged behind the United States in adopting and implementing eugenic measures. We can now see, from the perspective afforded by half a century of subsequent history, that Nazi medicine gave eugenics a bad name for only a relatively brief period. Today there are many who think we at last have adequate tools to begin to implement a serious program of eliminating "inferior" human beings and producing "superior" ones. Even if potential parents do not aspire to produce "superior" human beings, they are under considerable pressure to ensure that their children are without defects, are "perfect babies." In part, these pressures are the natural temptation to avoid the burdens of care for the handicapped, but such temptation is enormously amplified by the knowledge that those burdens are much easier to avoid today. In part, also, they are the pressures of cultural attitudes, assimilated by many Christians, to the child. For many it has become merely quaint to think of each child as a unique gift of God; children are more like planned acquisitions in our culture, acquisitions which should fit into our expectations about how our lives should go, about the ease and enjoyments that should characterize our lifestyle. A child who might threaten our ease may, if he or she is viewed as an acquisition, be thought of as a replaceable acquisition. And indeed genetic counselors will tell parents: you can terminate this pregnancy and try again for a "normal child." In addition to the strong appeal of convenience over justice in personal relationships, which is characteristic of our society, potential parents are subjected to pressures which have an economic motivation. This is evident both in systems of health care provision which are state controlled (as in the UK National Health Service) and in systems in which health care provision is insurance based. Thus in a report of the Royal College of Physicians on *Prenatal Diagnosis and Genetic Screening*[2] the College urged upon government the economic advantages of having a much more extensive program of genetic screening and selective abortion in the light of growing knowledge of gene-linked conditions. The case is very persuasive to a government intent on economizing in the provision of health care. When, just over two decades ago, the first nationwide program of prenatal diagnosis of neural tube defects and Down syndrome was introduced in the UK, it was done on the basis of pilot schemes in three regional health authorities, from which the health economists concluded

---

[2]Report of the Royal College of Physicians, *Prenatal Diagnosis and Genetic Screening: Community Service Implications* (London: Royal College of Physicians, 1989).

that the costs of schemes of prenatal diagnosis and selective abortion were amply justified by the considerable savings to be made of monies that would otherwise have been spent on the medical and social care of spina bifida and Down syndrome children.[3] Many clinicians in the National Health Service are influenced by the utilitarian cost/benefit mind-set of the health care administrators.

I have no very systematic picture of how economic pressures are brought to bear on parents who might have handicapped children in North and Central America and the Caribbean but cases, in the literature,[4] of insurance companies denying health insurance to people with genetic predispositions to disease seem to illustrate what one would expect in many private insurance-based health care systems. Potential parents of handicapped children, faced with the possibility of losing health care cover for their children, are obviously under significant economic pressure to avoid the birth of such children, especially when the insurance companies provide cover for "medically indicated" abortions.

In the rest of my paper, therefore, I want to concentrate on outlining some of the intellectual resources of the Catholic tradition which ought to help us stand up to the cultural, societal, and economic pressures which the growth of genetic knowledge is serving to intensify. We need to act as a people set apart in our approach to the begetting of children who may be handicapped.

There are three related fundamentals of the Catholic vision which should help us to counter the culture:

- our understanding of the nature of the human being— our anthropology;

- our understanding of human dignity; and

- our understanding of the relationship between marital commitment and the begetting of children.

---

[3]See S. Haggard and F.A. Carter, "Preventing the birth of infants with Down's syndrome: a cost-benefit analysis," *British Medical Journal* I (1976), 753–756; S. Haggard, F.A. Carter, and R.G. Milne, "Screening for spina bifida cystica: A cost-benefit analysis," *British Journal of Preventive and Social Medicine* 30 (1976): 40–53.

[4]See some examples in the position paper "Genetic Discrimination" prepared by the Council for Responsible Genetics (Cambridge, MA), January 2001 update. [http://www.gene-watch.org/programs/privacy/genetic-disc-position.html] (December 9, 2002)

You will be familiar with these but I think it would be helpful in the context of this workshop to bring out the way in which the Church's teaching on these three fundamentals provides us with an intellectual framework to confront one set of choices—reproductive choices—which developments in genetics are making more pressing for a not insignificant number of people who belong to your flocks. For the present, and indeed for the foreseeable future, choices for gene therapy (still less for germ-line gene therapy) are not where the main action is.

Christian Anthropology

Modern genetics, and the Human Genome Project in particular, have seemed to represent the achievements of a mechanistic view of human life and the human body in particular. By mechanism, in general, I mean the view that we can explain what a thing is by reference to the natural laws which govern the *parts* of the thing. If you want to understand something as simple as a bicycle pump, what you need to understand is how its parts are so fitted together that they take advantage of the natural laws which govern its parts—but it is the natural laws governing the parts (considered not *as parts but as objects with an independent nature of their own*) which ultimately explain how it works. A mechanistic understanding of any entity is, in principle, a reductionist understanding of that thing—one that seeks to get down to the most fundamental constituents of a thing and the laws governing those most fundamental constituents.

Mechanism, advanced as the correct approach to understanding living bodies, including the human body, goes back in the modern era to Descartes, but it took a long time to triumph in biology. The twentieth century by and large abandoned the tradition which still dominated nineteenth century biology of seeking to explain living entities by reference to the overall organization of the organism. That tradition was replaced by one for which life was to be understood in terms of genes and their replication. As a reductionist tradition, it seemed to have been wholly vindicated by the publication in 1953 of Watson and Crick's paper on the double helical structure of DNA. For Watson and Crick, there was a direct correspondence between the sequence of nucleotides in a gene and the sequence of amino acids in a protein. That was the central dogma of genetics in mid-century. Much subsequent work in the field of genetics has slowly served to subvert that dogma. Here is not the place to attempt to tell that story even were I competent to do so. But we should take note in passing of the implications of the central dogma for medicine.

It suggested that there would be a rather simple route to correlating abnormalities with defective or absent enzymes which in turn were to be explained by defects in genes. The central dogma has been subverted because research on gene function, as distinct from gene structure, has revealed an ever more complex picture. It has become increasingly clear that gene function is not explicable without reference to laws governing intra-cellular interactions, which in turn are not to be understood without reference to laws governing inter-cellular interactions, which in turn are not explicable except by reference to the roles of the organs to which the cells belong, which roles are not explicable except by reference to the life of the whole organism. Experimental work in genetics, is then, forcing a retreat from a reductionist to an organismic explanatory framework.[5]

Let me just mention as an aside here, that what is emerging as the hierarchical complexity of organisms is part of the explanation for what is called the therapeutic gap between knowledge of genetic structure and our ability to screen for genetic defects, on the one hand, and, on the other, possession of medical treatments. I do not wish to seem to belittle the great advances which have been made in understanding the role of genetic mutations in the causation of many diseases. The links between a number of conditions and mutations in particular genes have been demonstrated, most clearly with the single-gene disorders (such as Tay-Sachs, Huntington's Chorea, cystic fibrosis, and the thalassemias). But even here a good deal remains, I believe, unexplained about the processes linking the state of the gene to the onset of disease. This is the background to the observation by Professor David Weatherall, Director of the Institute of Molecular Medicine at the University of Oxford, that: "Transferring genes into a new environment and enticing them to ... do their job, *with all the sophisticated regulatory mechanisms that are involved*, has, so far, proved too difficult a task for molecular geneticists."[6] And it is because of this view on the present state of play in the development of gene thera-

---

[5]This is a clear inference one can make from Evelyn Fox Keller's *The Century of the Gene* (Cambridge, MA, and London, England: Harvard University Press, 2000).

[6]D.J. Weatherall, "How Much Has Genetics Helped?" *Times Literary Supplement*, January 30, 1998, 4–5. Emphasis added.

pies that it seems reasonable to assume that gene therapy is not where the action is for the present. These observations on the "therapeutic gap" are, however, an aside to our main theme.

What I have principally wanted to emphasize here is the way reductionism has proved inadequate in that part of biology which has most aspired to be reductionist—namely genetics— and the progressive recognition of the need to refer to the life of the organism as a whole in order to make adequate sense of the functioning of genes in the organism. Now it is precisely to make intelligible the ongoing life of the organism as a whole that Aristotle employed the concept of the "soul."

Christian anthropology has employed a broadly Aristotelian concept of the "soul" in definitively articulating that part of the Church's teaching about man which emphasizes the unity of the human person. As Pope John Paul II has recalled in *Veritatis splendor*, the "rational soul is *per se et essentialiter* the form of [the human person's] body." He goes on to explain: "The spiritual and immortal soul is the principle of unity of the human being, whereby it exists as a whole—*corpore et anima unus*— as a person. These definitions not only point out that the body, which has been promised the resurrection, will also share in glory. They also remind us that reason and free will are linked with all the bodily and sense faculties."[7] In other words a Christian anthropology is radically opposed to any anthropology which understands the human body in essentially mechanistic terms and sees it as something external to the human person. A dualism between person and body is characteristic of much writing and thinking in the field of bioethics; historically it goes back to Descartes. Characteristically, on such a view, personal existence is understood to be a function of already developed, exercisable abilities for thought and choice. Hence the readiness, in our day, to treat as disposable those who have not yet developed these abilities or who have lost them.

Christian Anthropology and Human Dignity

St. Thomas Aquinas, proceeding from the definition of a person as "an individual substance of a rational nature," ex-

---

[7]Pope John Paul II, Encyclical Letter *Veritatis splendor* (Vatican City: Libreria Editrice Vaticana 1993), n.48, with footnote references to the Ecumenical Council of Vienne *Fidei catholicae*, DS 902, and the Fifth Lateran Ecumenical Council, Bull *Apostolici regiminis*, DS 1440. Original emphasis.

plained that human beings are constituted as individual substances of rational nature in virtue of having rational souls. The rational soul is the "form and life-long actuality"[8] which gives dynamic unity to the complex material organisms we are and to the expression of our various powers (vegetative, animal and intellectual) in multifarious activities. St. Thomas says: "From the essence of the soul flow powers which are essentially different ... but which are all united in the soul's essence as in a root."[9] The powers of the soul, as wholly undeveloped radical capacities, are given to each individual at the beginning of his or her existence. This is the basis of the natural dignity which belongs to every human being.[10]

An important part of St. Thomas's teaching is his insistence that there is only one soul which makes us to be what we are. And one implication of this insistence is that the human body shares in the dignity of our rational nature in the sense that, as long as we are living human bodies, our bodiliness is intrinsic to what gives us natural dignity. Because the human body is integral to our personal existence, the characteristic goods of our animality, such as life, health, and the transmission of life, are integral to our fulfilment as persons and are to be protected, promoted, and respected as such.

The truth that the principle of life of every living human being is a rational soul and the truth that each human being has, in consequence, a radical capacity to develop rational abilities, whether or not he or she has done or will do, provide grounds for us to hold all human beings to be equal in dignity. But these truths, on their own, provide us with an incomplete sense of human dignity. What we need to add to these truths are the truths that each human soul is "immediately" created by God[11] "in his own image" and that the proper destiny of each human being is eternal life with God.

---

[8]John Finnis, *Aquinas: Moral, Political and Legal Theory* (Oxford: Oxford University Press, 1998), 178.

[9]Thomas Aquinas, *Commentary on the Second Book of the Sentences of Peter Lombard*, d.26, q.1, a.4c.

[10]This and the previous paragraph draw heavily on a much more extensive discussion of anthropology and dignity, Luke Gormally, "Human Dignity: The Christian View and the Secularist View" in *The Culture of Life: Foundations and Dimensions*, eds. J. Vial Correa and E. Sgreccia (Vatican City: Libreria Editrice Vaticana 2002), 52–66.

[11]See *Catechism of the Catholic Church*, n.366 and references there.

Pope John Paul II has recalled these truths, fundamental to a culture of life, in his encyclical *Evangelium vitae*:

> The life which God gives man is quite different from the life of all other living creatures, inasmuch as man, although formed from the dust of the earth (cf. Gn 2:7, 3:19; Jb 34:15; Ps 103:14; 104:29), *is a manifestation of God in the world, a sign of his presence, a trace of his glory* (cf. Gn 1:26-27; Ps 8:6). This is what St. Irenaeus of Lyon wanted to emphasize in his celebrated definition: "Man, living man, is the glory of God." Man has been given *a sublime dignity*, based on the intimate bond which unites him to his Creator: in man there shines forth a reflection of God Himself....
>
> In the biblical narrative, the difference between man and other creatures is shown above all by the fact that only the creation of man is presented as the result of a special decision on the part of God, a deliberation to establish *a particular and specific bond with the Creator*: "Let us make man in our image, after our likeness" (Gn 1:26).... *The life* which God offers to man *is a gift by which God shares something of himself with his creature*....
>
> The life which God bestows upon man is much more than mere existence in time. It is a drive towards fullness of life; *it is the seed of an existence which transcends the very limits of time*: "For God created man for incorruption, and made him in the image of his own eternity" (Wis 2:23) (n. 34, original emphasis).

The life of each human being, his or her bodily existence, is a "gift from above." We do not *have* the gift of life; there is no pre-existent "me" to be a recipient and subsequent possessor of that gift. Rather *we* are what are given in God's giving of the gift of life: our personal existence is a direct gift from above, a gift the destiny of which is entry into the communion and friendship of God's own Trinitarian life.[12]

Marital Commitment and a Right Relationship to Children in Their Origin

Because it is fundamental to the nature of our existence that we *are* gift, it is fundamental to the recognition of our dignity that we should be received and accepted as gift. And since

---

[12]These thoughts are in part inspired by Robert Spaemann, "On the anthropology of the Encyclical *Evangelium Vitae*" in *Evangelium Vitae: Five Years of Confrontation with the Society* [sic], eds. J. Vial Correa and E. Sgreccia (Vatican City: Libreria Editrice Vaticana, 2001), 437–451.

what is most basic to shaping a sense of our identity which is consistent with our dignity is our begetting, gestation, birth, and early rearing, it is of crucial importance that we have parents who are so related to each other that they are disposed to receive us as "gifts from above."

Now marriage is designed precisely to dispose parents to receive their children as gifts from above, destined for friendship with God. As Pope Paul VI said in *Humanae vitae*:

> It is false to think that ... marriage results from chance or from the blind course of natural forces. Rather, God the Creator wisely and providentially established marriage with the intent that He might achieve his own design of love through men. Therefore, through mutual self-giving, which is unique and exclusive to them, spouses seek a communion of persons. Through this communion, the spouses perfect each other so that they might share with God the task of procreating and educating new living beings.[13]

Integral to God's plan for bringing human beings into being is that it should take place in and through a relationship of unreserved self-giving love. Husband and wife are fitting cooperators with God in the transmission of human life just to the extent that they have an unreserved love for each other. For the generosity of their love is the vehicle, so to speak, through which the child encounters the love of God. More profoundly, the unreserved character of that love is the condition of the child being accepted for who he or she is, a gift of God whose dignity is only adequately acknowledged in unconditional acceptance. Unconditional acceptance of the child as child and gift of God is born, under grace, out of unconditional acceptance of each other by husband and wife.

As I have already remarked, the new genetics is having its largest scale impact on the lives of ordinary people through programs of screening, either pre-natal screening for fetal abnormality, which standardly goes along with the offer of abortion if the unborn child is suffering from an abnormality, or pre-implantation screening, intended to detect, where there is a known risk of genetic defects, which in vitro embryos have genetic defects and which do not, so that only the latter are placed in the womb. Other proposed techniques for avoiding the implantation of embryos with genetic defects rely on extra-cor-

---

[13]Pope Paul VI, *Humanae vitae* (1968), trans. Janet Smith, in *Humanae Vitae: A Generation Later*, Janet Smith (Washington, D.C.: The Catholic University of America Press, 1991), 269–295, at n. 8.

poreal conception, possibly with the use of gametes or parts of gametes from persons other than the putative parents. The last mentioned techniques are clearly inconsistent with the requirement that a child should be begotten of the bodily self-giving of husband and wife, since some gametal material derives from persons other than the husband and wife. But given that a standard approach to avoiding the risk of transmitting genetic defects is in vitro fertilization followed by pre-implantation screening, it is important that we should have a clear grasp of why this is inconsistent with recognition of the dignity of the child; recall that the dignity of the child is essentially linked to the truth that he or she is a gift of God.

The reason was stated as follows in the 1987 instruction from the Congregation for the Doctrine of the Faith, *Donum vitae, On Respect for HumanLlife in its Origin and on the Dignity of Procreation*:

> [F]ertilization is licitly sought when it is the result of a "conjugal act which is of the kind that is suitable for the generation of children to which marriage is ordered by its nature and by which spouses become one flesh." But from the moral point of view procreation is deprived of its proper perfection when it is not desired as the fruit of the conjugal act, that is to say, of the specific act of the spouses' union.[14]

Why is it said that these techniques deprive procreation of its proper perfection, in other words result in it failing to be the kind of activity it should be?

The basis of the teaching is exactly the same as the basis of the teaching in regard to contraception: the inseparability of the unitive and procreative meanings of sexual intercourse. What the Church says about the begetting of children is that the behavior which serves to bring them into being should be behavior expressive of the mutual self-giving, the unity in one flesh, of their parents. Why? Because human generation needs to have as its human cause the kind of act which is expressive of the unreserved self-giving of husband and wife. If a child's coming-to-be has its human origin in such an act, then the

---

[14]Congregation for the Doctrine of the Faith, *Donum vitae, Instruction on Respect for Human Life in its Origin and on the Dignity of Procreation* (Vatican City: Vatican Polyglot Press, 1987), II, B.4, quoting from Pope Pius XII, "Discourse to those taking part in the Second Naples World Congress on Fertility and Human Sterility," May 19, 1956: *AAS* 48 (1956), 470.

child enters the relationship of husband and wife as the fruit of *unreserved* self-giving. It is precisely and only that status which is adequate to the dignity of the child: for truly unreserved self-giving carries with it a commitment to unreserved acceptance of the fruit of that self-giving. The dignity of the child is only adequately recognized in the acceptance and cherishing of him just as he is. The disposition to that acceptance is protected precisely by the teaching that what human beings do to bring a child into being should be an action of the parents which is also expressive of their two-in-one-flesh unity.

The rejection of that teaching undermines at root the disposition to accept children as they are *given to us*, and as, therefore, the persons they are. That rejection has had incalculable consequences, both in the fields of clinical practice and biomedical research, and more broadly in society in people's attitudes to children. The development of the "reproductive technologies" assumes the acceptability of separating the generation of children from the sexual act expressive of the two-in-one-flesh unity of their parents. In consequence, human beings generated in vitro have come to *seem* to be *the manipulable products of technical expertise* rather than the fruit of unconditional self-giving. In vitro embryos, rather like mass-produced objects, are subject to quality control and discarded if deemed unfit for implantation; others are generated solely for experimentation which is destructive in character. There is a body of work in genetic engineering which looks to a future in which designer babies will be produced to parental specifications. The root of these developments, which are so deeply at variance with what is required for recognition of the dignity of children, lies in the rejection of the teaching that children should be begotten as the fruit of marital intercourse which is expressive of the unity of the parents.

This teaching, as I mentioned earlier, is the obverse of the teaching that it is only intercourse which is of the generative kind (i.e., intercourse not deliberately rendered sterile) which is truly unitive, in other words, truly marital. Since a majority of Catholics, often with the encouragement of priests, have rejected that teaching, it is not surprising that few of them can make any sense of the Church's teaching about reproductive technologies. Yet that teaching is indispensable to openness to and acceptance of the child as a gift from God. And if Catholics are not, through marital chastity, deeply disposed to children as the gifts of God, they will be terribly vulnerable to eugenic ideology and eugenic practices to which developments in genetics are giving such a powerful impetus.

I have made negative eugenic procedures associated with the new genetics a central theme because, for the present at least, these procedures are likely to be the most conspicuous manifestations in the lives of ordinary people of the new genetics. I have urged that marital chastity is indispensable to openness to and acceptance of our children as gifts from God.

Catholics will not have the dispositions they need to honor and cherish each child as a gift from God, and to resist the intensified eugenic pressures of our age, if they have not had a formation in chastity. There is no more relevant response to contemporary eugenics than the effort to help people see that what the Church teaches about chastity, and about contraception in particular, are liberating truths made livable by God's unfailing love for us. In so far as we are made chaste through the power of that love we are well-disposed to accept and love our children for what they are—gifts of God.

MORAL IMPLICATIONS OF GENETIC ALTERATION IN HUMANS

MARILYN E. COORS

Genetic alteration offers the promise of wonderful thera-peutic benefits accompanied by equally significant moral chal-lenges. The Pope specifically addressed the issue of genetic al-teration in humans in 1982 while addressing the Pontifical Academy of Sciences. He endorsed the attempt to cure heri-table genetic disorders, such as cystic fibrosis, through the transfer of genes that ameliorate the crippling effects of chro-mosomic and genetic disease. He extended this approval to ge-netic interventions with embryonic and neonatal life, given that all such manipulations "be subject to moral principles and val-ues which respect and realize in its fullness the dignity of man."[1]

[1]Pope John Paul II, "The Ethics of Genetic Manipulation: John Paul II to Medical Associations," Origins 13 (November 17, 1983): 386–389, at 387.

The National Conference of Catholic Bishops also acknowledged the tremendous benefits of genetic medicine, while expressing the attendant risks associated with genetic research. Given the potential of genetic science to modify all forms of life, the bishops asserted the need to fully explore the intended ends of genetic intervention.[2]

Philosophers also emphasize the mandate for caution as they foresee the fruition of genetic science at the advent of the third millennium. They debate the discrepancy between man's aptitude for knowledge and the ability to predict the consequences of that knowledge. Unforeseeable consequences could result from the ability of scientists to effect direct intervention into the human genome. The limits of knowledge imposed by human finitude are inescapable. Science can tell us a great deal about the prevention, treatment, or cure of genetic disorders, but it cannot help us distinguish between right and wrong. It can tell us what can be done, but it cannot guide us as to what should be done. Science cannot dictate moral criteria for attaining the good for humankind, but it must inform moral deliberation concerning the new genetics.

The first portion of this presentation addresses the science that relates to the moral implications of genetic alteration. An overview of the potential applications of human genetic alteration follows. Finally, a summary of the directives of the Holy Father on the ethics of genetic alteration provides the framework upon which to assess the applications of the genetic revolution.

The Science

The announcement in July 2000 of the completion of the mapping and sequencing of the human genome stimulated tremendous excitement and rational apprehension as well.[3] The excitement centers on the promise of new and improved therapies for genetic disorders; the apprehension focuses on the concern that the power of DNA exceeds what humans can

---

[2]National Conference of Catholic Bishops, "Statement on Recombinant DNA Research" (May 2, 1977), in vol. 4, *Pastoral Letters of the United States Catholic Bishops: 1975–1983*, ed. Hugh J. Nolan (Washington, D.C.: United States Catholic Conference, 1984), 200–204.

[3]"The Human Genome," *Nature* 409.6822 (entire issue February 15, 2001): 745–964; "The Human Genome," *Science* 291.5507 (entire issue February 16, 2001): 1145–1434.

manage wisely. The complexity and profundity of the information encoded in the human genome substantiates both reactions.

The human genome is the total aggregate of genetic material contained in the twenty-three pairs of chromosomes in each human cell; some refer to the genome as the "book of life." The genome contains approximately thirty-five to fifty thousand genes. A gene is the basic functional unit of heredity; a gene is to heredity what a sentence is to a book. Genes consist of the macromolecule deoxyribonucleic acid (DNA). The DNA molecule is comprised of four nucleotide bases, a sugar (deoxyribose), and a phosphate. Nucleotide bases in the genome are analogous to letters in the alphabet. The four nucleotide bases are to genes what the twenty-six letters are to words. The sequence of nucleotide bases in a gene determines its function, just as the sequence of letters in a word ascertains its meaning. A change in the arrangement of bases results in a dysfunctional gene or a mutation. To continue the analogy, a different arrangement of the same letters can result in a different or meaningless word. For example, transposition of the letters in the verb "pare" produces the noun "pear." The two words have different grammatical functions and unrelated meanings and will disrupt the meaning of a sentence if they are interchanged.

Every gene provides information that instructs cellular machinery to produce the proteins necessary for human life. The normal functioning of all of the genes in concert, each producing precise amounts of protein at specific times in exact locations, is necessary for the development and maintenance of human life. When a change occurs in a sequence of DNA bases that direct the manufacture of a normal protein, the cell produces a nonfunctional protein, no protein at all, or the incorrect amount of protein. A DNA miscoding sequence resembles "typing mistakes that would be expected of a secretary transcribing a biochemical manuscript some three billion characters long."[4] The mistake, absence, or dysfunction of a necessary protein can cause disease. Some examples of disorders caused by a DNA mutation are sickle cell anemia, cystic fibrosis, Huntington disease, type two diabetes, and many cancers.

---

[4]Constance Engelkind, "The Human Genome Project Exposed: A Glimpse of Promise, Predicament, and Impact on Practice," *Oncology Nursing Forum* 22 (Supplement 1995): 27–34, at 30.

As the scientific gaze penetrates the microcosm of each individual cell and gains new understanding, the complex interrelationship of cellular activity and biological life become readily apparent. There is an order and rhythm in the genetic dance of life that is now visible but will never be totally comprehensible. As the chromosomes divide and replicate, one witnesses the creative function of a God who is orderly, sustaining, and aesthetic. The genetic processes are never contingent or chaotic, albeit sometimes faulty. The action and interaction of the chromosomes that influence the development and maintenance of life display a beauty and complexity that reflects the intellect of the Creator of the Universe. When scientific investigation encounters the core of human life, understanding evolves into mystery and awe.

It is now possible for human actions to alter the course of existence with limited grasp of the ramifications of those actions. In order to participate responsibly in the moral debate surrounding human genetic alteration, the clergy and laity must understand the scientific facts and moral implications of this technology.

Genetic Alteration

Genetic alteration is the transfer of genes to cells to replace dysfunctional genes with functional genes; the technique can be used to correct errors in the genetic process or to improve normally functioning processes. Genetic alteration occurs in two types of cells: 1) somatic cells (body cells) or any cells lacking the potential to become reproductive cells; and 2) gametes (reproductive cells) or early embryos. A transferred gene inserted into an early embryo assimilates into virtually all of the cells of the ensuing fetus, together with germ cells. The technical obstacles to the safe and effective alteration of gametes or early embryonic cells remain daunting. One major impediment is the questionable capacity of transfer mechanisms to deliver the transgenes without disrupting the delicate levels, timing, and/or distribution of gene expression. The development of new transfer mechanisms, such as artificial chromosomes, may overcome this hurdle, but the effect of interjecting a different number of chromosomal segments into the human genome is still unknown. To return to the linguistic metaphor, the notion of enriching a book by splicing in parts of another book is illogical because in books, as in people, ignoring the context is invariably counterproductive. However, scientists claim that the introduction of an artificial chromosome into the genome is more like adding a preface to a book than interjecting a chapter at random.

The distinction between somatic cells and germ-line cells is crucial to understanding the important bioethical, philosophical, and moral issues involved in genetic alteration. Somatic-cell gene alteration is the transfer of a functional gene into body cells. The intent is that the transferred gene will operate in the cell to restore function or express a given trait. Genetically altered somatic cells convey transferred genes to daughter cells upon cell division. However, they do not pass on transferred gene(s) to future generations. Ethically, somatic cell gene alteration is an extension of standard medical practice into the arena of otherwise intractable genetic disorders. It is not significantly different from organ donation, blood transfusion, or many other medical procedures. Organ transplantation involves an outside donor, the informed consent of the parties involved, and affects only the individual receiving treatment. The procedure results in significant differences for the post-surgical patient whose future existence is contingent upon the technological intervention. Somatic-cell gene alteration is equivalent to transplantation in that the same medical and ethical parameters apply. Medical interventions are ethically acceptable when they comply with the principles of autonomy (informed consent), beneficence (do good and avoid harm), and justice (fairness).

It is also technically feasible to deliver functional genes into germ-line cells. The technology utilized in germ-line gene transfer is analogous to somatic cell gene transfer with one significant difference. Genetic changes in germ-line cells are permanently encoded in the reproductive cells of future persons and ultimately become part of the human gene pool. The dividing line between somatic-cell and germ-cell alteration is not always distinct. Gene transfers specifically intended to alter somatic cells could inadvertently affect germ cells, passing those changes on to future generations. For example, adult males produce gametes throughout their lives. Therefore, genetic alteration intended to affect only somatic cells could unintentionally affect the production of gametes. Germ-line alteration raises distinct ethical issues in that its effects are cumulative, specific, and impact generations as yet unborn. This is a departure from any former medical procedure because germ-line alteration entails intergenerational consequences that were heretofore unfathomable. Additionally, potential future persons are not included in present ethical considerations of obligations or responsibilities.

The intended outcomes of somatic-cell and germ-line alteration are often differentiated according to purpose: therapy or enhancement. Therapeutic genetic alterations are procedures intended to restore health through the correction of dysfunctional gene(s). In contrast, enhancement genetic alterations are procedures intended to improve traits or function above normal standards. The purpose of genetic enhancement is to produce more desirable qualities than would be the result of natural reproduction. Genetic enhancements that objectify human life are morally problematic based on the virtues of justice and wisdom and the potential to result in new forms of discrimination in society.[5]

The distinction between gene therapy and gene enhancement can also present problems. There are many cases in which the intent of genetic alteration is obvious, but there are some in which the distinction becomes problematic. For example, Huntington disease is a late onset fatal neurodegenerative disorder. A genetic alteration to correct the gene that causes Huntington disease is clearly therapeutic. However, genetic alterations that correct genes that predispose to common complex disorders, such as heart disease or autoimmune disorders, are less clear. It is possible that genetic alteration techniques could eliminate heart disease, thereby increasing life expectancy by several decades. Is this therapy or enhancement? Moreover, what if the autoimmune genes were identified and the "altered" individual could function under conditions that were detrimental to others. How would this alteration be classified? And where would we draw the line in the alteration of height in a short-statured male: 4'4", 5'4", or 6'4"? It is difficult to assess intent in some cases. There is a natural progression from clearly therapeutic objectives to enhancement purposes that leaves a "gray" area that is obscure. I raise this problem not to confuse the issue, but to encourage the evaluation of new advances in genetic medicine on an ongoing basis not only at the technical level but also at the philosophical and moral levels.

---

[5]Pope John Paul II, "Regard for Human Dignity Must Guide Genetic Intervention," *Health Progress* 65 (January 1984): 47. See also Mark S. Frankel and Audrey R. Chapman, "Inheritable Genetic Modifications: Assessing Scientific, Ethical, Religious, and Policy Issues," (Washington, D.C.: American Association for the Advancement of Science, 2000). [http://www.aaas.org/spp/dspp/sfrl/germline/report.pdf] (August 6, 2002).

Applications of Genetic Alteration

The application of genetic alteration in humans remains ten to twenty years in the future, but, at present, the pressure to produce healthy children through assisted reproductive technology anticipates potential uses. Given the present state of reproductive technology, this translates into genetic testing and selection of those conceptions that are deemed acceptable and disposal of "imperfect" or "surplus" embryos.[6] Both of these techniques work very well if the goal is to avoid genetic disease. If the goal is to protect the sanctity of human life, both techniques are immoral. The unborn child is required to pass a genetic test in order to be born. As pressure increases and the withdrawal of insurance coverage for "preventable" disorders looms, genetic alteration in utero could provide a moral alternative for those opposed to the destruction of human life. Theoretically, it will become possible to test a fetus and alter cells in utero once a delivery system for the gene transfer is perfected. For many developmental disorders such as Down syndrome or other multi-organ disorders, early detection and genetic alteration could prevent irreversible damage to the fetus before it occurs.

In addition, new techniques for analyzing DNA, called microarray chips, will impact the perception of what it means to produce healthy children. DNA chips permit an individual's DNA to be tested for multiple, or eventually all, genetic disorders simultaneously. Chip technology is presently available; only the identification of the function of the approximately thirty-five thousand human genes remains. This means that in the fairly near future DNA testing of an embryo, a fetus, or a newborn will predict risk for all future disorders. Once scientists can analyze the genome thoroughly, there will be no such thing as a perfect embryo or fetus, since every conception carries defective genes. So, as science can test for more and more disorders, there may come a point when genetic alteration provides a justifiable option, technically as well as morally, for those who believe in the sanctify of every human life.

This explanation should not be interpreted as a defense of germ-line alteration. It merely presents a picture of the medical landscape in the near future and analyzes the ways in which this technology might be used in a moral fashion.

---

[6]Ted Peters, "In Search of the Perfect Child: Genetic Testing and Selective Abortion," *Christian Century* 113 (October 30, 1996): 1034–1037.

Arguments for and against Germ-Line Alteration

Germ-line alteration is a radically new procedure that is not encompassed within the present moral domain. Previously, ethics was conducted among contemporaries. The potential now looms for human action to impact our descendants more directly than was possible heretofore. Therefore, the ethical domain must expand to include those future persons impacted by present actions. The scientific and ethical complexities inherent in the evaluation of germ-line alteration are obvious. Along with the promise of tremendous benefit, there exists the potential for abuse. The successful demonstration of the efficacy of germ-line alteration technology in mice in 1994 at the University of Pennsylvania generated ethical concerns earlier than most geneticists or ethicists expected. Sheldon Krimsky commented:

> Human genetic modification has begun without a clear consensus on where the moral boundary lines should be placed to insure that the technology of human genetic engineering is not abused.[7]

Currently germ-line alteration in humans is considered unethical by most in the scientific and religious communities.[8]

However, there are some voices endorsing the utilization of germ-line alteration techniques. A synopsis of the four ethical arguments endorsing germ-line alteration is as follows. The first is medical necessity. It encompasses the obligation to rectify developmental disorders or those that affect multiple systems but are not amenable to somatic-cell treatment. Germ-line modification could remedy these otherwise untreatable afflictions permanently. The second argument is medical efficiency. Germ-line gene alteration is more efficient than the repetitive somatic-cell gene alteration over successive generations because somatic-cell modification involves its own risks, as discussed above. Thus, it would be safer and more cost effective to eliminate a disorder in both the present and future persons in one procedure. Third, germ-line alteration offers an alternative to repeated prenatal diagnosis and selective abortion in families afflicted with genetic disease. As the demand to produce healthy children

---

[7]Sheldon Krimsky, "Human Gene Therapy: Must We Know Where to Stop before We Start?" *Human Gene Therapy* 1.2 (Summer 1990): 171–173, at 171.

[8]Frankel and Chapman, "Inheritable Genetic Modifications."

increases, this argument gains credibility. The final justification holds that the medical profession should pursue and implement the best available techniques to prevent or treat genetic disease.

The ethical arguments opposing germ-line alteration in humans are numerous and profound. To summarize the debate, it centers on the issues of potential clinical risks (scientific uncertainty), inviolability of the human genome, the social implications of genetic information, and the limit of human wisdom. All medical treatment involves uncertainty about outcomes, but germ line alteration entails a heightened level of complexity and longevity. Moreover, the inviolability of the human genome encompasses a notion of the sacred that regards the value of the human genome as innate, not commercial. Recent attempts to commercialize the genome raise the social concerns surrounding genetic information to a heightened level. Issues such as privacy, informed consent, genetic discrimination, reduction of life to a DNA code, and strain on family relationships take on new meaning as more and more genes are linked to disorders and traits. Finally, the argument stressing the limits of human wisdom requires the acknowledgement of the enormity of our ability to act contrasted with the paucity of our ability to evaluate the ramifications of that action. The writings of John Paul II provide the moral direction to address the issues surrounding genetic alteration well in advance of the clinical application of this technology.

The Writings of John Paul II

The moral direction of the Pope reflects the reality of an order in the created universe deserving of respect. He invites humans to participate in the work of the Creator with the statement: "The researcher follows God's design. God willed man to be the king of creation."[9] The Book of Genesis positions man as the steward of the earth, the only creature endowed with reason and the discernment of good and evil. But the account of creation also designates boundaries. The Pope requires that genetic alteration of human life respects the order demonstrated by nature and the essence of humankind as created by God.

John Paul II cites three conditions that must be met in order for human genetic alteration to be morally acceptable. They

---

[9]John Paul II, "The Ethics of Genetic Manipulation," 389.

are as follows. First, the alteration must respect the biological integrity of every human being as a unity composed of body and mind. Second, embryonic life must be accorded the basic rights inherent in all humans. Third, the manipulation of the genome may not aim at the creation of new or different groups of people. These three requisites delineate the limits beyond which genetic alterations of the human genome are proscribed.

He underscores the unity of the human being in order to emphasize the dignity that is due to a creature made in the "image of God" (Gn 1:28). Scientists transgress the limits of wisdom when genetic alteration attempts to "modify the genetic store." The Pope's concern is that genetic alterations could include created differences that "provoke fresh marginalization" in our world.[10] This refers to the enhancement or degradation of human traits that compromise the integrity of the human person. For the Pope, the intent underlying the genetic alteration is of primary importance.

> The fundamental attitudes inspiring the intervention we refer to should not derive from a racist, materialist mentality aimed at human happiness that is really reductive. Man's dignity transcends his biological condition.[11]

Human transcendence, dignity and freedom, must be protected from technological assault.

In further specifying the parameters of the moral limits of human genetic alteration, John Paul II takes a position in accord with the therapy/enhancement distinction. He endorses therapeutic interventions as long as the harmful effects do not exceed the anticipated benefit.

> A strictly therapeutic intervention, having the objective of healing various maladies—such as those which stem from chromosomic deficiencies—will be considered in principle as desirable, provided that it tends to promotion of man's personal well-being, without harming his integrity or worsening his life conditions.[12]

He does not condemn genetic enhancements that are not strictly therapeutic but "aimed at improving the human biological condition." He requires that two conditions be fulfilled given the profound moral significance of improving traits rather than correcting a defect. The intervention must not impair the

---

[10]John Paul II, "Regard for Human Dignity," 47.

[11]Ibid.

[12]John Paul II, "The Ethics of Genetic Manipulation," 388.

origin of human life, and an intervention must respect the dignity of humankind and the "common biological nature" that forms the basis of human liberty.[13]

The somatic-cell/germ-line distinction took precedent in a more recent address. Following the same parameters of the "transcendental vocation" of humankind and the "incomparable dignity" of the human persons that constantly ground his principles, the Pope endorsed medical treatments that ameliorate "fatal hereditary pathologies." In doing so he endorsed the possible elimination of those pathologies: "By acting on the subject's unhealthy genes, it will also be possible to prevent the recurrence of genetic diseases and their transmission."[14] This statement indicates that germ-line alteration aimed at the prevention or transmission of genetic disease that does not interfere with the origin of human life could be considered.

He explains that a strictly scientific explanation of human life negates freedom and clashes with the "irrefutable evidence" that the human spirit cannot be reduced to an object but consistently remains the author of our actions and beliefs.[15] His comments reveal a foreboding engendered by the reductionist potential of the new genetics: "The ability to establish the genetic map should not lead to reducing the subject to his genetic inheritance."[16] The treatment of genetic illness must not be isolated to the technical problems posed by the treatment, but the "patient in all his dimensions" warrants primary consideration.[17] The term "dimensions" refers to the unity of the human person, the wholeness of the "affective, intellectual, and spiritual functions." In his definition of dimensions John Paul includes personal relationships, circumstances, and history.

Jesus Christ is the incarnation of the moral law. As such, he will always be the norm for embodied human life.[18] The book of Colossians states: "He is the image of the invisible God, the

---

[13]Ibid.

[14]Pope John Paul II, "The Human Person—Beginning and End of Scientific Research, Address of Pope John Paul II to the Pontifical Academy of Sciences (October 28, 1994)," *Pope Speaks* 40 (March/April 1995): 80–84, at 81.

[15]Ibid., 82.

[16]Ibid.

[17]John Paul II, "Regard for Human Dignity," 47.

[18]John Paul II, "Holy Father on 'Human Genome Project,'" *L'Osservatore Romano*, December 1, 1993, 3.

firstborn over all creation" (Col 1:15). Thus, Jesus is the example of moral action and the model for human physiological life. This is not to imply a normative gender, class, race, or appearance but, rather, a standard for the crucial cognitive, biological, spiritual, emotional, and relational elements that are genetically determined and shared by all human beings, regardless of condition. Ultimately, the unity of qualities that He exemplified must circumscribe the moral limits for human genetic alteration.

GENETIC COUNSELING: CHALLENGES FROM THE NEW GENETICS

SUSAN SCHMERLER

The first draft of the Human Genome Project has been completed. One predicted impact of this project in medicine has been a shift in the focus from diagnosing and treating disease to identifying an individual's predispositions in order to minimize or prevent disease altogether.[1] Even a limited impact on health care, such as the ability to tailor drugs for people with inherited differences in sensitivity,[2] could be far reaching.

Most ethical challenges raised by the new genetics are not new or different from those issues raised in other areas of health care. The Human Genome Project and the information

---

[1]F.S. Collins, "Shattuck Lecture: Medical and societal consequences of the Human Genome Project," *New England Journal of Medicine* 341.1 (July 1, 1999): 28–37.

[2]N.A. Holtzman and T.M. Marteau, "Will genetics revolutionize medicine?" *New England Journal of Medicine* 343.2 (July 13, 2000): 141–144.

generated by it cause us to look at how we weigh the ethical principles involved with these issues.

People are generally interested in knowing the probability of their getting a disease and what can be done to either prevent the disease or to improve the outcome. But when risks have been determined to be genetic rather than to be caused by some environmental agent, there is a perception that the risk is less amenable to change.[3]

We are going to look at how the profession of genetic counseling can respond to some of those issues. To begin, we will examine the philosophic basis of the field of genetic counseling, then discuss some of the challenges that have begun to emerge as a consequence of the Human Genome Project. Medical advances cannot influence people's health if they are fearful, whether the fear is due to a lack of knowledge or to an absence of essential public policy. Genetic counselors have a valuable contribution to make in this area.

Genetic Counseling Practice

Definition

When a family has an inherited disorder, for example, the muscular dystrophy syndrome myotonic dystrophy, where do they learn that it is caused by an autosomal dominant gene that can be passed from generation to generation? Who will address the family's emotional reaction to the diagnosis, the fears for the other members of the family, the future for affected and unaffected family members? In the late 1960s, Virginia Apgar recognized that there was a need for a professional who could work with families, interpreting and translating medical information that had been presented by physicians during the evaluation and diagnosis of an individual's health problems, that is, in times of great stress. The profession of genetic counseling as we know it today developed to meet that need.

I started working as a genetic counselor about the time the most often quoted definition of genetic counseling was published:

> Genetic counseling is a communication process, which deals with the human problems associated with the occurrence or risk of occurrence of a genetic disorder in a family. This process involves an attempt by one or more

---

[3]Ibid.

appropriately trained persons to help the individual or family to; (1) comprehend the medical facts including the diagnosis, probable course of the disorder, and the available management; (2) appreciate the way heredity contributes to the disorder and the risk of recurrence in specified relatives; (3) understand the alternatives for dealing with the risk of recurrence; (4) choose a course of action which seems to them appropriate in view of their risk, their family goals, and their ethical and religious standards; and act in accordance with that decision; and (5) to make the best possible adjustment to the disorder in an affected family member and/or to the risk of recurrence of that disorder.[4]

This definition includes the goals of client education, assistance with decision making, and aiding in psychosocial adjustment. Today, over twenty-five years later, most counselors and clients agree that information provision and increased knowledge are not only the goals, but also the outcomes of genetic counseling.[5] The provision of genetic information by genetic counselors can be considered a medical intervention. Although it is broader in focus than some medical treatments, it has the potential of forming the basis for many lifetime choices.

There are different approaches to providing genetic counseling. One approach is the *teaching* model, where facts and information are conveyed to equip the client to make a rational decision. For the family with myotonic dystrophy, the autosomal dominant inheritance pattern, the fifty/fifty risk for every child of the affected person, the equal risk for boys and girls, are all part of the information they need to plan their future. The *counseling* model provides time, space, and reinforcement of the client's own competence and capacity for autonomy. It allows the individual with myotonic dystrophy to think about his future, about what he may or may not be able to do as a career or for recreation.[6] For example, maybe a career in computers would be more successful than a career as an athlete. The goal of both models is a "good" decision, that is, a decision

---

[4]F.C. Fraser, "Genetic counseling," *American Journal of Human Genetics* 26 (1974): 636–661.

[5]B.A. Bernhardt, B.B. Biesecker, and C.L. Mastromarino, "Goals, benefits, and outcomes of genetic counseling: Client and genetic counselor assessment," *American Journal of Medical Genetics* 94.3 (September 18, 2000): 189–197.

[6]S. Kessler, "Psychological aspects of genetic counseling VII: Thoughts on directiveness," *Journal of Genetic Counseling* 1.1 (1992): 9–17.

that is fully informed, well reasoned, and balances the client's goals, values, and circumstances with the moral and social implications of that decision.[7] Ethical considerations are essential to these decisions and need to be integrated into the process.[8]

Philosophy and Ethos

The philosophy and ethos of genetic counseling reflects the profession's development as a specialty within medicine, so that the principle-based ethics that guide medical practice also influence genetic counseling. The principles of autonomy, nonmaleficence, beneficence, and justice shape much of the approach of genetic counselors. Because genetic counselors respect the individual's autonomy, the values of privacy, confidentiality, and informed consent are given a high priority.[9] Because of this influence, the values of the voluntary use of services, equal access to services, full disclosure, and nondirective counseling are emphasized.

Clients must be free to come to their own decisions and even make their own mistakes.[10] The utilization of genetic services is the first step in the process, and the decision to make use of genetic counseling ideally needs to be voluntary. Pressure by others, no matter how good the reason, whether from physicians, family, or commercial interests, can interfere with the decision-making process. Genetic counselors strive to make genetic services available to all who need and want them. We also strive to disclose all the relevant information to the client while providing education about diagnosis and related issues.[11]

---

[7]M.T. White, "Making Responsible Decisions: An Interpretive Ethic for Genetic Decision-Making," *Hastings Center Report* 29.1 (January/February 1999): 14–21.

[8]C. Grady, "Ethics and genetic testing," *Advances in Internal Medicine* 44 (1999): 389–411.

[9]National Society of Genetic Counselors, "National Society of Genetic Counselors Code of Ethics," *Journal of Genetic Counseling* 1.1 (1992): 41–43.

[10]White, "Making Responsible Decisions."

[11]B.A. Fine, "The Evolution of Nondirectiveness in Genetic Counseling and the Implications of the Human Genome Project," in *Prescribing Our Future: Ethical Challenges in Genetic Counseling*, eds. D.M. Bartels, B.S. LeRoy, and A.L. Caplan (New York: Aldine de Gruyter, 1993), 101–117.

The principle-based approach to problems that arise in genetic counseling has been complemented by an ethic of care that is based on the understanding that relationships are an important factor in moral decision making.[12] A nondirective-counseling approach is the norm.[13] Although at first glance it may not seem apparent, nondirective counseling is an active strategy, a way of thinking about the professional/client relationship in which at each step the professional attempts to evoke the client's competence and ability of self-direction.[14] Genetic professionals are at times directive, providing guidance in some areas, such as with regard to behaviors that are considered therapeutically beneficial, e.g., taking extra folic acid during pregnancy, avoiding drugs that can cause birth defects, or undergoing screening of at-risk populations.[15]

Information about an individual's family history, genetic status, diagnosis, or risk for genetic disease has the potential to be stigmatizing, and can result in discrimination. Client confidentiality has been given high priority by the genetics community for that reason.[16]

Genetic Counseling Challenges from Genetic Medicine

Genetic technologies and genetic testing raise important questions about identity, differences, social tolerance, and even the meaning of disease.

As more genes are identified, there will be an increase in the use of genetic tests and the information that they generate.[17] What genetic counselors address today that we did not discuss twenty-five years ago are the ethical dilemmas that may arise as a result of genetic testing, especially those issues that can impact insurance coverage or employment.

---

[12]J.L. Benkendorf et al., "An explication of the National Society of Genetic Counselors (NSGC) Code of Ethics," *Journal of Genetic Counseling* 1.1 (1992): 31–40.

[13]J.R. Sorenson, "Genetic Counseling: Values That Have Mattered," in *Prescribing Our Future.*

[14]Kessler, "Psychological aspects of genetic counseling VII."

[15]M.B. Mahowald, M.S. Verp, and R.R. Anderson, "Genetic counseling: Clinical and ethical challenges," *Annual Review of Genetics* 32 (1998): 547–559.

[16]Fine, "The Evolution of Nondirectiveness."

[17]Grady, "Ethics and genetic testing."

Informed Consent

There is a wide consensus that the decision to have or not to have a genetic test should be made by the individual. The process of informed consent is essential to that decision, but is being challenged in the era of the genome.

Careful assessment of the risks and benefits of a test for every disorder should be made. The client's assessment is going to vary with the problem being tested for, with the quality and reliability of the assay, with the state of knowledge regarding the interpretation of the defects identified, as well as with the available choices of action.

Multiplex testing, doing many genetic tests on a single sample, is now available. One spot of blood from a newborn can be tested for over twenty biochemical disorders. The knowledge we can get from the laboratory is coming faster than thoughtful, well-considered decisions as to how and when to use this knowledge to improve health.[18] One approach that has been suggested to help manage the number of genetic tests, for example, is the use of a generic consent. This is a "one consent fits all" proposal to cover several tests at one time. Instead of full disclosure, explaining each test, for example, for Tay-Sachs, Gaucher, Niemann-Pick, the consent would be for the heritage panel of Jewish genetic diseases. This process obviously fails to provide adequate disclosure, and informed consent as we define it is impossible.[19]

For consent to be valid, it must be voluntary. Pressures from family members, health care providers, researchers, and others have become more significant in light of the changing perception of the meaning the test results have to each of these groups.[20] For example, 1) under the influence of virtue ethics, the provider's judgment was not often questioned, but since the principle of autonomy became prominent, the individual's judgment directs her choices; 2) under the principle of beneficence, the physician/patient relationship was strictly private, but the principle of justice suggests that the family may have a claim on a family member's genetic information; 3) at one time medical researchers searched for information mainly for the ben-

---

[18]Ibid.

[19]S. Elias and G.J. Annas, "Generic consent for genetic screening," *New England Journal of Medicine* 330.22 (June 2, 1994): 1611–1613.

[20]Grady, "Ethics and genetic testing."

efit of others, while today there is a rush to patent genes and to form biotech companies; and 4) employers and health insurers, although restricted in many U.S. states, claim a right to have the results of genetic information and to use it.

Although voluntary participation is a goal for genetic testing, genetic tests are routinely done on all newborns, many without the explicit consent of their parents. The benefits of these tests have been determined to outweigh the risks, and only conditions, such as thyroid problems, that can be treated (and where serious complications can be avoided) are included. Testing under these conditions is considered acceptable care.

When the question of doing genetic tests for children arises, the medical community agrees that in general children should not have genetic tests unless there is a direct medical benefit to that child from testing. In a family that may have a condition such as Huntington disease that does not develop until adulthood, testing would not be done on a child. But if the family has multiple colon polyps that become cancerous, a condition in which the polyps can be found in childhood, then testing would be very beneficial to the child and should be done.[21]

Advances in genetics have also precipitated the need for genetic counseling in all areas of primary care and medical specialties (especially oncology). We can anticipate that nongeneticists will likely provide the bulk of counseling. A very great concern is that nongeneticist caregivers have not been prepared for that role. Medical specialists already find it difficult to absorb and retain genetic information outside their own field of expertise.[22] The task of training and educating these care providers is already underway. In this pursuit the genetic counselor's role is that of teacher and expert.

Confidentiality

Traditional patient care is a one-to-one interaction. How should we accommodate the enormous amount of information that is becoming available to families?[23] Several techniques have been suggested: group counseling, Internet information sites, written material. Each of these methods presents a chal-

---

[21]ASHG/ACMG Report, "Points to consider: ethical, legal, and psychological implications of genetic testing in children and adolescents," *American Journal of Human Genetics* 57.5 (November 1995): 1233–1241.

[22]Mahowald, Verp, and Anderson, "Genetic counseling."

[23]Ibid.

lenge to the values derived from the principle of respect for patient autonomy. When groups are used, the challenge is especially to confidentiality. The presence of third parties can inhibit both the patient and the provider and increases the possibility of a loss of confidentiality. Use of the Internet, where there is no quality control in place, and the use of written materials (brochures, pamphlets) and educational videos, further challenge the counselor/client relationship and weaken those values we strive to uphold in the counseling relationship.

It is also becoming increasingly more difficult to protect the privacy of genetic information. We all agree that the tested individual should receive the information and should make decisions about future disclosures. With the societal shift from emphasizing autonomy to justice, the claims of the family to what is shared information are becoming more urgent. Although most practitioners encourage individuals to share information with those relatives for whom it has importance, dilemmas arise when the client refuses. Also, samples taken for genetic testing can be stored for years, used for other purposes, and shared with other laboratories. DNA databanks have already been established by the federal government for the future identification of war dead, and by law enforcement agencies for criminal identification.

Genetic testing has been found to reveal undesired or unexpected results. The most commonly used illustration is of nonpaternity. Another is of susceptibility to unanticipated diseases, such as the risk for Alzheimer disease that was found through the ApoE4 testing done for cardiac patients. Careful counseling prior to testing can help avoid the repercussions of some of these ancillary findings, but as the science progresses, counselors may have to discuss this in general terms.

Genetic counseling has been offered by a group of individuals who are welltrained in genetics. As we mentioned in relation to informed consent, with the increasing demand for the service, the patient's introduction to genetic testing will become the responsibility of primary care physicians. Practicing medicine without thinking about genetics will become impossible within the next few years. Members of the public, including health care providers, generally have little understanding of the basics of genetics, probabilities, and risks. Education will be a large part of the role of the genetic counselor in the coming years.

Therapy

The use of the information becoming available from the Human Genome Project is opening multiple avenues for the prevention and/or treatment of disease while it raises ethical and social questions.

Some diseases can be treated by altering the various genes that are in certain cells of the body. Somatic-cell gene therapy has not moved out of the laboratory and into clinical practice yet. But once that happens, such a success will stimulate interest in research on germ-line cells that would enable the correction of the disease to be passed on to future generations.

We are all aware of the death of Jesse Gelsinger, the young man with the biochemical disorder ornithine transcarbamylase (OTC) deficiency, who was participating in experimental gene therapy. The family knew the risks from that gene therapy itself. A posttreatment vector mutation occurred, which can happen especially when the vector is integrated into the host genomic DNA, resulting in vector-driven tumor formation. This mutation is possible . Although this may have been understood intellectually, the possible outcome was not fully comprehended.

There may be unexpected harms from placing new genes into either somatic or germ-line cells. The new gene could disrupt the function of other genes near the site of insertion. If the techniques should prove to be of limited use in curing disease, the focus of research could shift to genetic enhancement. How will genetic counselors approach the question of moving beyond therapy to the use of these techniques for genetic enhancement, that is supplying a characteristic that a parent might want in a child. This does not involve the treatment or prevention of disease. How should genetic counselors respond when parents request such therapies?

I think it is unlikely that genetic engineering will develop from therapeutic techniques to determine traits of children, or even change the probability of developmental improvements. These interventions will have uncertainties and the final expression of the genetic enhancement will be dependent on the environment and on the developmental history of the child.[24] We need to keep in mind that differences in social structure, lifestyle, and environment account for much larger proportions

---

[24]W. Gardner, "Can Human Genetic Enhancement Be Prohibited?" *Journal of Medicine and Philosophy* 20 (1995): 65–84.

of disease than genetic differences.[25] Yet this subject must be considered and addressed in the public arena.

Prevention

Genetic heritage binds us vertically to biological parents and children, and horizontally to siblings. When an individual seeks information about her own genetic profile, something is learned about her parents, children, and siblings. Client and physician need to begin to rethink obligations in a "clan-based" way.

Information will have an effect on, and decisions may be required from, other members of the family who may be concerned. Family members have a clear interest in the diagnosis of genetic disease, carrier screening, and increased susceptibility. It is interesting to note that when asked who they would trust with their genetic information, patients who were at risk for genetic disorders with serious complications said: ninety-six percent would tell partners, ninety-five percent would tell other family members at risk if the information was perceived as useful to these relatives, and eighty-nine percent would tell members of the clergy.[26] What is the responsibility of the professional when the individual does not want to share this information? We have to weigh our ethical obligations to the patient with those to families and society. When does society have a greater claim on the information than the patient's claim to confidentiality?

As tests move from the research laboratory into clinical use, what is the role of the professional? We will have to continue to warn of the genetic risk that is faced; that has not changed. We will have to disclose availability of testing, and provide information regarding changes of lifestyle or aspects of the environment. There will be many former patients who may benefit from new tests or treatment possibilities. Will we have to recontact these patients as new tests are developed?

Concern for justice goes beyond individuals to consider the common good. We do not want genetic testing to be used against an individual seeking employment. But on the other hand, screening employees with periodic examination for chromosome

---

[25]Holtzman and Marteau, "Will genetics revolutionize medicine?"

[26]J.O. Weiss, C. Kozma, and E.V. Laphan, "Whom would you trust with your genetic information?" *American Journal of Human Genetics* 61.4 (October 1997): A24.

or DNA damage can pinpoint hazardous substances and lead to measures to reduce exposure to them. One-time testing to determine an individual's genetically-based susceptibilities could lead to protection of employees from injury- or illness-inducing exposures. Employers may have legitimate concerns about productivity, absenteeism, and turnovers. However, we want to prevent health insurance cost containment obtained through the elimination of very high cost users.

Public Policy

Society has to avoid the temptation to focus attention on the fascinating but remote possibility of human cloning, for example, and address the more fundamental questions of what constraints should be placed on obtaining genetic knowledge, what limitations should be placed on the use of that knowledge, and what comprehensive ethical and public policy framework will be used in dealing with the implications of the new genetics.[27]

Inequities in health care deny some people the choice of genetic tests, while inadequate descriptions of the nature, purpose, and implications of testing may hinder the autonomy of others. Genetic tests may not identify some people at risk and they may misclassify others.[28] While yielding opportunities for clinical intervention, genetic screening may have significant adverse effects on a person's image, social standing, employment prospects, and insurability. Genetics is not neutral in its impact on different groups.[29] The principle of justice demands efforts to reduce the gap between those who are advantaged and those who are not.

An important question that will face society is to what extent, if any, genetic disorders or predispositions should provide a basis for determining access to certain social goods, such as employment or insurance. There is a suggestion that health and life insurers are already using genetic information to deny coverage or reimbursement. The fear of being denied insurance deters some individuals from being tested for genetic disorders or susceptibilities. Genetic test results can also be used

---

[27]M.C. Kaveny, "Genetics and the Future of American Law and Policy," *Concilium* 1998.2 (April 1998): 57–72.

[28]G. Burnside, "Genetic Medicine in the Next Five Years," *Health Progress* 80.5 (September–October 1999): 43–45, 48.

[29]Mahowald, Verp, and Anderson, "Genetic counseling."

to deny or limit employment of individuals by employers who are concerned about the costs of health insurance as well as the possibility of genetic susceptibility to illness caused by workplace toxins.[30]

Correlating genetic test results with the occurrence of disease is costly and takes a long time for tests done early in life for late-onset disorders (e.g., Alzheimer disease). The cost could deter test development.[31] On the other hand, genetic medicine can be expected to lower the cost of health care. Finding out a person's individualized disease predisposition long before he becomes ill will lead to his health being preserved. This is often less expensive than treating the illness. Also, quantifying to some degree a woman's genetic predisposition to breast cancer, for example, could lead to different strategies for mammography screening. That could lead to a more rational use of health care dollars. This can be a way to overcome some of the barriers to access.[32]

For people who are asked to participate in genetic research, privacy can become an issue. They themselves are protected by the rules that apply to human subjects, but what protects the data that is gathered, the tissues that are saved in tissue banks, the need of other researchers for samples? This issue can only be addressed on a public policy basis, and inroads have been made by several states in protecting genetic privacy.

The family nature of genetic information raises other public policy issues. Should a person be required to learn about her genotype if it would benefit other family members? A positive test result for one person, such as a grandmother who is found to have a breast cancer gene, can provide information that is useful to many other family members who would then be considered at a high risk to also be gene carriers. Do we want, as a society, to create a duty for individuals who have tested positive for a DNA mutation to tell their families or should that duty go to the genetic counselor?

Finally, to the extent that genetic medicine is becoming big business, do we as a society want to create parameters for this new enterprise?

---

[30]Grady, "Ethics and genetic testing."

[31]Burnside, "Genetic Medicine in the Next Five Years."

[32]Ibid.

Goals and Challenges

The goals of genetic counseling include providing objective and balanced education of clients and supporting the client's independent decision making. The choice of whether to undergo testing and the management of genetic information are the client's to make. It is generally recognized that facts alone are insufficient for making these decisions and for coping with the implications. The psychosocial and ethnocultural context in which information is received and perceived as relevant to the family or to the individual is crucial.[33] Genetic counselors present information in a way that is useful for the individual.

We have reviewed the challenges facing both patients and professionals from the influx of new opportunities for testing, therapy, and research generated by the Human Genome Project. We are also provided with an opportunity to influence the direction of the discussions and the establishment of significant public policies. We must take advantage of these opportunities.

---

[33]Fine, "The Evolution of Nondirectiveness."

Genetic Counseling: Genetic Predispositions, Privacy Rights, and Access to Genetic Information

Albert Moraczewski, o.p.

Ethical Considerations
in the Light of Church Teachings

In the previous chapter, Dr. Susan Schmerler presents an excellent overview of genetic counseling. She rightly includes the philosophy which undergirds the practice of genetic counseling and briefly outlines some important ethical dimensions of that process: informed consent and confidentiality, therapy and prevention, and then concludes by noting some public policy issues.

I propose in this brief note to sketch out briefly some of the important moral considerations from the perspective of the Catholic moral tradition. I shall include three cases to illustrate some of the difficult moral and prudential decisions that one can encounter and to suggest a process by which these can be dealt with in a reasonable manner in accordance with the Church's moral tradition.

For the moral position I am taking here, the point of departure is that every human being is made in the image of God, and more specifically, called to be remade in the image of Jesus Christ, true God and true Man. In that truth resides the intrinsic dignity of each human person. As a consequence, each person conceived in this world is due appropriate respect, regardless of their stage of development, their medical condition, or social status. And conversely, for the same reason, each human being has the moral responsibility to respect the other. If a being has the moral right to be respected, then it has, in turn, the responsibility to respect the dignity of others, with due respect to the limitations that being's current condition may impose, e.g., being in a coma.

The hubris of many individuals and groups, encouraged by the technological prowess which has rapidly developed during the past few decades, is moving them to propose public policies aimed at preventing the birth of children whose genetic profile does not meet the standards they have set. They claim that these genetically challenged individuals will not only have an unhappy life (because of the alleged poor quality of life), but also will be a great drain on public resources. These "social engineers" are the driving force behind a renewed eugenics program.

The above having been said, what about the current practice of genetic counseling? If a person or couple resort to genetic counseling, what sort of ethical issues will they be likely to face? The following three cases will illustrate some of the moral issues which individuals and couples may face in dealing with information acquired as a result of a genetic diagnosis. The names of the individuals and some of the details have been modified so as to protect the privacy of those concerned.

Three Case Studies

Tay-Sachs Disease

Tay-Sachs disease is an autosomal recessive disease (i.e., both parents have to be carriers of the defective gene) due to an inherited deficiency of an enzyme (hexosaminidase A) necessary for the proper metabolism of certain lipids (i.e., fats). The enzyme is encoded by a gene, *HEXA*, located on the long arm of chromosome 15. This genetic deficiency leads ultimately to developmental retardation, paralysis, mental deterioration, blindness, and finally death at about age three.

A couple, Mr. and Mrs. T.S., are both in their middle fifties. The first of the two daughters, born in 1950, was for the first eight months of life (apparently) completely normal. When she failed to sit up the parents were concerned and consulted their pediatrician. The results of tests done at age twelve months suggested that the child was afflicted with Tay-Sachs disease. Over the following thirty-two months there was a gradual deterioration: loss of normal motor function, weakness, inability to feed, blindness, frequent seizures, and during the last months she entered a vegetative state and died at the age of three and a half years. The family suffered immensely as they watched the progressive and inevitable deterioration. The second daughter developed normally and married at the age of twenty-three. Genetic testing of the daughter was positive for Tay-Sachs; she was a carrier; her husband also tested positive. The wife having become pregnant, the couple decided to have prenatal diagnosis (amniocentesis) done in order to determine whether the child had the disease (there was a twenty-five percent chance that she was afflicted). They had decided after much deliberation, and vigorous input from her parents, to have the child aborted if the tests were positive for Tay-Sachs. The necessary tests were done and came up positive for Tay-Sachs.

May the couple request an abortion in order to spare the child a few years of pain and discomfort, and to spare his parents the heavy emotional and financial burden of caring for a severely handicapped child?

No, the wife may not have an abortion. The Catholic Church's moral tradition prohibits direct abortion for any reason whatever. The deliberate killing of an innocent human life, regardless of age, stage of development, or medical condition is intrinsically evil. "[T]he Church teaches that 'there exists acts which *per se* and in themselves, independently of circumstances, are always seriously wrong by reason of their object.'"[1] The Pope continues by noting that Vatican Council II lists a number of examples of acts which are intrinsically evil. This list includes abortion. In *Evangelium vitae,* Pope John Paul makes the solemn statement

> Therefore, by the authority which Christ conferred upon
> Peter and his successors, and in communion with the bish-

---

[1] Pope John Paul II, *Veritatis splendor,* n. 80, quoting *Reconciliatio et paenitentia,* n. 17.

ops of the Catholic Church, *I confirm that the direct and voluntary killing of an innocent human being is always gravely immoral* (n. 57, original emphasis).

How one pastorally communicates this moral conclusion to the couple in question and counsels them is a pastoral challenge. One has to balance carefully compassion (in the manner of communication) with the preservation of moral truth (the content of the communication).

Huntington Disease

Huntington disease is inherited as a dominant condition, so that the possession of a single gene for Huntington disease will usually result in the person having that condition. Hence any child of an affected parent will have a one in two (fifty percent) chance of inheriting that disease. It is a neurological disease which includes uncontrolled muscular movements, and progresses through dementia to death over a ten- to twenty-year span. The symptoms of the disease, however, do not become manifest until about age thirty-five. Over twenty thousand persons in the United States have Huntington disease. The *HD* gene is located on the short arm of chromosome 4.

A couple, Harry and Dorothy, plan to marry and have several children, at least four or five. But Harry has just learned from a phone call that his mother died of Huntington disease. While Harry knew that his mother, living in a remote rural area of the country, had some sort of a muscular disorder, he did not know the nature of the disorder until her death. After going, secretly, to a genetic counselor, he learned that he had a fifty percent chance of having the same disorder and that each child that he sired would also have a one in two chance of inheriting Huntington disease. He was now faced with several moral problems: 1) Should he tell his intended wife? 2) Should they marry but mutually agree not to have children? 3) And, if they marry, should Harry have a vasectomy and/or his wife have a hysterectomy to make sure of not begetting a child?

Is Harry morally obligated to share with his fiancée the information about his condition before they marry? If they agree before marriage not to have any children, would that invalidate the marriage? In light of the horrendous nature of this disease and the high probability, one in two, of a child conceived by that couple having Huntington disease, may the husband (or wife) undergo direct sterilization?

Since there is a fifty percent probability that Harry is a carrier of Huntington disease, there would be a significant impact on his future wife since she would be burdened with his care when the symptoms became manifest. In addition it would affect greatly their decision about having any children. Consequently it is a matter of justice that she be informed before the marriage take place.

Canon 1055 stipulates that "The matrimonial covenant ... is by its nature ordered toward the good of the spouses and the procreation and education of offspring."[2] While the actual procreating of a child is not required, "there must nevertheless be an openness to procreation by all who choose this sacrament."[3] Thus, if a couple freely, explicitly, and absolutely, for whatever reason, exclude the procreating of children in their marriage, that union would be invalid.

They may not submit to direct sterilization, whether permanent or temporary, because direct contraceptive sterilization is intrinsically evil. Pope Paul VI's encyclical *Humanae vitae* (1968) clearly stated the Church's teaching "that each and every marriage act (*quilibet matrimonii usus*) must remain open to the transmission of life."[4] Any conjugal act performed when the husband has had a vasectomy or the wife an hysterectomy would, *in se*, by deliberate choice, not be open to the transmission of life. Even though the couple has the right (and perhaps a moral obligation) to avoid bringing a child into the world who has a lethal disease, they do not have the right to use any means whatever. The couple have the option to observe complete abstinence, which for most couples would be a heroic measure, but they also have the option to use Natural Family Planning (NFP). Unfortunately, NFP has gotten a "bad rap" in the popular press. However, it is a very effective method of limiting births when employed properly and consistently.[5] Hence, if they limit intercourse to those infertile times which nature provides, they

---

[2]James A. Coriden, Thomas J. Green, and Donald E. Heintschel, eds., *The Code of Canon Law: A Text and Commentary* (New York: Paulist Press, 1985), 740.

[3]Ibid.,741.

[4]Pope Paul VI, *Humanae vitae* (Boston: St. Paul Books & Media, 1968), n. 11.

[5]See R.E.J. Ryder, "'Natural family planning': Effective birth control supported by the Catholic Church," *British Medical Journal* 307 (September 18, 1993): 723–726. See also Thomas Hilgers, "Family Planning Issues: NFP, Norplant, Uterine Isolation," in *The Interaction of Catholic Bioethics and Secular Society*, ed. Russell E. Smith (Braintree, MA: The Pope John Center, 1992), 213–230.

would both avoid having a child with a lethal condition and be doing so in a morally acceptable way. It is understood that there is an objectively grave reason for the use of NFP and, in addition, that the couple would nonetheless freely and lovingly accept any child conceived and born, *praeter intentionem.*

Congenital Hypophosphatasia Lethalis

This is an inheritable disease transmitted as an autosomal recessive condition (i.e., both parents have to pass on the defective gene) is due to the lack of an enzyme, alkaline phosphatase, and results in the diffuse lack of calcium deposition in the bones. This leads to dwarfism and bone deformities, and this form terminates in death within the first few years of life.

Mr. and Mrs. C.H. were married for three years when they had a child, Pam, who died at the age of four as result of having inherited the disorder congenital hypophosphotasia lethalis. Their physician has advised the couple that if they elect to have another child they should do so by IVF so that a preimplantation genetic diagnosis could be performed on the embryo. This procedure would give them the opportunity to accept or reject an afflicted embryo.

At least two principle issues arise when the couple want another child:

1) May the couple request preimplantation genetic diagnosis in order to determine whether the newly conceived child has this condition?

2) If the couple really want a child and forgoes the preimplantation genetic diagnosis, may they proceed to take the natural steps for begetting a child notwithstanding the fact that there is one in four (twenty-five percent) chance that the child will have that disease?

Preimplantation genetic diagnosis may not be requested since the procedure requires that conception take place as a result of in vitro fertilization (IVF) in order to have ready access to the embryo for the analysis. In 1984 the Church declared that IVF was not morally acceptable because it violated the dignity and sacredness of human procreation.[6]

There is no clear moral obligation to avoid procreation. A prudential judgment must be made. One must recall in this

---

[6]See *Donum vitae,* II, B.5.

situation that there is a three out of four (seventy-five percent) chance that a normal child (that is, free of this inherited condition) will result. Given those odds, it is not certain that the couple would be under a moral obligation not to bring a child into this world who may be burdened with this serious genetic disease. Perhaps the question should be raised as to why this couple want another child in light of the death of the previous child as a result of this disease. Is the reason selfish? Is it for religious reasons? Is it for the good of the child, e.g., potential for eternal life? Is it for the good of an older but single child? In addition, is this couple capable—financially, emotionally—of rearing such a child? What is their "track record" with regard to the previous child who died of the disease? Have they considered the possibility of adoption?

If the answers to these questions indicate that the couple is not acting out of selfish reasons and is capable of accepting and properly rearing an afflicted child, it would be imprudent to lay on their shoulders the burden of not seeking another child.

These three cases illustrate different modes in the transmission of inheritable disorders and some of the moral concerns that arise when these cases are viewed from the perspective of the Catholic moral tradition. Dr. Schmerler raised some of the ethical concerns which could arise from a secular ethic.[7] Of course, these are also concerns of the Church, even if the analyses would proceed along other paths. And sometimes the moral conclusions would be the same.

---

[7]See Susan Schmerler, "Genetic Counseling: Challenges from the New Genetics," in the previous chapter, 131–143.

HUMAN DEVELOPMENT—THE LONG AND SHORT OF IT

JOHN M. OPITZ, M.D.

The *annus mirabilis* 2000 celebrated not only a Jubilee year, but also the centenary of quantum physics, of Mendelian genetics, and the bicentenary of morphology. During the extraordinary scientific (mostly biologic) watershed of 1800, powerful causal currents entered the Western contemplation of Nature, one historical, the other developmental, to provide a modern harmony or counterpoint to the Aristotelian, or more appropriately, neo-Platonic view of nature which dominated "Continental" and also New World biology under the guise of "*Naturphilosophie*" until the beginning of the nineteenth century. Considering the compelling allegiance of most Western biolo-

---

I am most grateful to the Foundation of the Primary Children's Medical Center for a generous grant in support of the International Clinical Genetics Research and Consultation Program, Division of Medical Genetics, Department of Pediatrics, University of Utah, and to Mrs. Sandie Ramos for secretarial collaboration.

gists to that philosophical perspective during the early nineteenth century, it is truly astounding how much they contributed by way of concrete biological, anatomical, histological, embryological, paleontologic, cytological, and ultimately, genetical facts and data to that gigantic body of knowledge that constitutes modern biology.

"Preformation," that philosophical view postulating the origin of development to the moment of creation whereby all subsequent generations were preconceived in either Eve (the "ovists") or Adam (the "spermists"), yielded reluctantly to the epigenetic point of view, initially verbalized by Aristotle and William Harvey,[1] whereby each living being arises not from a preformed complete but miniature creature from sperm or egg, but from an initially incomplete formless mass of tissue or group of cells, the result of the junction of one male and one female germ cell, gradually assuming form, or coming into being, through a process called development, "informed" by genes, the origin of which can now, reliably, be projected back to 3.5–4.0 billion years, one billion years or so after the coming-into-being of the earth.

Now, two hundred years after the beginning of morphology and a veritable revolution in science which has come closer than ever to the goal of uniting chemistry, physics, and biology, we are approaching an understanding of the relationship between development and evolution, still based, as it was a century ago, on the analysis of *homology*, i.e., the molecular basis of the form and formation of living beings. Genetics, having arisen out of morphology over a century ago, is now universally understood as the science of the *causal* analysis of development (quite in the spirit of Gregor Mendel's use of the term "*die bildungsfähigen Elemente*") and recently was reunited with morphology through the mediacy of molecular biology.

During the nineteenth and early twentieth centuries the sciences of embryology[2] and experimental embryology[3] gave us the knowledge of *what* happens during development and of the

---

[1]G. Harvei (William Harvey), *Exercitationes de generatione animalium* (Londini apud Octavianum Pulleyn/Amstelodami apud Ludovicum Elzevirium, 1651).

[2]C. Darwin (1859), *The Origin of Species by Means of Natural Selection, or the Preservation of Favored Races in the Struggle for Life* (New York: The Modern Library, 1998).

[3]W. Roux, *Programm und Forschungsmethoden der Entwickelungsmechanik der Organizmen* (Leipzig: Engelmann, 1897).

genetics that explain *how* it happens. In this manner we have learned that the theory of descent *must* be correct since all mammals/vertebrates, indeed also invertebrates, share the same developmental and genetic control mechanisms, a fact which makes it obvious that *all* organisms on this earth, plants and animals, constitute a single family of life, totally interdependent and exceedingly fragile and vulnerable, with many weak members under threat of extinction.

The success of the human species is all the more astonishing given that most (some ninety percent[4]) potential human beings die spontaneously prenatally due to major genetic (chromosomal) defects. This fact, and the extraordinary complexity of the interdependence of mother, placenta, and fetus, makes it imperative to provide the survivors of this massive prenatal selection process with the best conditions to assure optimal outcome for mother and infant. The robustness of the survivors depends substantially also on maternal health, nutrition, and peace of mind, and may be impaired by maternal stress, malnutrition, smoking, alcohol and drug intake, parental consanguinity in inbred populations, advanced maternal age, and multiparity. Thus, it would seem most appropriate that all of us supporting families, pregnant women, and children, spiritually and medically, join forces in a synergistic effort to assure the best condition for the development of a joyful and committed spirit in a healthy body.

Two Hundred Years of Morphology

What Johann Wolfgang von Goethe (1796) and Johann Friedrich Burdach (1800)[5] designated "*Morphologie*" was rapidly recognized as an exciting new approach to biology, incorporating for the first time developmental and historical perspectives previously lacking in static anatomy and comparative anatomy. When Goethe said "*Gestaltenlehre ist Verwandlungslehre*"[6] he in-

---

[4]J.M. Opitz, "The Farber Lecture: Prenatal and perinatal death: The future of developmental pathology," *Journal of Pediatric Pathology* 7 (1987): 363–394.

[5]J.M. Opitz and A. Rauch, "Von der befruchteten Eizelle zum Menschen: genetische Defekte als Schlüssel zum Verständnis der menschlichen Ontogenese" (From the fertilized egg cell to the human being: Genetic defects as key to an understanding of human ontogeny) in *Gene, Neurone, Qubits und Co: Unsere Welten der Information*, eds. D. Ganten et al. (Berlin: Gesellschaft Deutscher Naturforscher und Ärzte; Leipzig: S. Hirzel Verlag, 1999), 237–254.

[6]"The study of form is study of its change."

timated that the study of (static) form was the simultaneous study of the formation (development) of the individual organisms and their transformation over time from ancestral to present species.

Morphological studies at the beginning of the nineteenth century began with tremendous enthusiasm for:

- the study of extinct, older forms of life (paleontology, i.e., the pioneering work by Georges Cuvier of 1812 and 1817 on ancient fossils[7];

- the study of prenatal or preadult stages of many animals (later called "embryology" by Charles Darwin in *The Origin of Species* in 1859, and "ontogeny" or "ontogenesis" by Ernst Haeckel in 1886[8]); and

- the astonishing similarity in structure, and the development of structure, between animals, initially referred to as "analogous" but later as both analogous (if serving a similar function but representing a different structure, e.g., a butterfly and a bat wing) and "homologous" (if similar in structure even if function is dissimilar, e.g., a mole paw for digging, a whale flipper for swimming, and a bat forelimb for flying).

Homology in structure (and development of structure) was one of the most powerful arguments presented to Darwin by early nineteenth century morphologists and was summarized to the effect that homology of structure in different species represented homology of embryonic processes by virtue of descent, with modification, from a common ancestor with prototypic developmental plan.[9]

Transformation of species through modification of developmental processes (the argument of "descent") was well formulated before Darwin. What Darwin added was a *causal* postulation—namely, the agency of natural selection. Darwin did not use the word evolution urged on him later by Herbert Spencer,

---

[7]D. Outram, *Georges Cuvier: Vocation, Science, and Authority in Post-Revolutionary France* (Manchester: Manchester University Press, 1984).

[8]E. Haeckel, quoted in *A Glossary of Genetics and Cytogenetics*, R. Rieger, A. Michaelis, and M.M. Green, 3rd ed. (Berlin: Springer, 1968).

[9]S.F. Gilbert, *Developmental Biology*, 6th ed., (Sunderland, MA: Sinauer, 2000).

but the last word in *The Origin of Species* is "evolved" used in its modern sense. In 1866 Haeckel coined the term "phylogeny" or "phylogenesis" to refer to the development of species.[10]

Most of the progress in morphology in the nineteenth century was made by physicians who also studied plants and animals, normal and abnormal, and early on contributed to a study of teratology, which was then understood as the study of *all* malformations (not only those environmentally caused). Thus, the definition of "morphology" in the nineteenth century was "the science of the form, formation, transformation, and malformation of living beings."

As the nineteenth century progressed, morphology, to an ever increasing extent, became a *causal* science, natural selection being postulated as the cause of evolution, glaciers as causes of landscape and mountain formation, and successive climatic catastrophes as causes of periodic extinction and fossil formation of large numbers of formerly living organisms. Indeed, the adjective "genetic" was used widely in nineteenth-century science and biology to refer to causal factors, long before William Bateson[11] made a noun and a science of it by adding an "s." Like Darwin, who sought for the causes of evolution in a morphological context, so also Mendel was a child of nineteenth century morphology, seeking to find the causes of development (i.e., of specific attributes of living organisms such as peas and bees) through systematic breeding efforts. To his deep astonishment and gratification, he discovered not only the causes of development (which, true to his morphological tradition he called the "*bildungsfähigen Elemente*," i.e., elements with form-giving or morphogenetic potential[12]), but also their behavior in transmission from one generation to the next during the process that would be called *meiosis* in the next century. Darwin desperately needed a causal hypothesis for the origin and transmission of form but seems never to have cut the pages of the reprint of the work that Mendel sent him.

When, during the latter part of the century, the experimental embryologists began to investigate developmental mechanisms,

[10]E. Haeckel, *Generelle Morphologie der Organismen* (Berlin: Reiner, 1866).

[11]W. Bateson, *The Methods and Scope of Genetics* (Cambridge: Cambridge University Press, 1908).

[12]G. Mendel, "Versuche über Pflanzen-Hybriden," *Verhandlungen des Naturforschenden Vereines in Brünn* 4 (1866): 3–47.

the manifesto of the founder of that branch of biology, Wilhelm Roux, stated: "The causal method of investigation is experimental. Certainty in causal deduction can only come from experiment, either from 'artificial' or from 'nature's experiment,' such as variation, monstrosity, or other pathological phenomena."[13] Note the equal emphasis in that statement on *experimental* manipulations of the embryo and on naturally occurring *dysmorphogenetic phenomena.*

One of the saddest legacies of the early passion for genetics which swamped biology after the rediscovery of Mendel's laws a century ago was its fierce denigration of the scientific culture and tradition of form and formation, replacing these concepts with the term "phenotype" which was studied as an attribute of the gene, and no longer as a matter worthy of study *per se.* This created in North America, and to some extent also in Europe, an ever deepening schism in biology, such that genetics was breeding, transmission, and cytogenetic work, and morphology "mere" medical school anatomy, histology, and pathology.

Thus, it came to pass that geneticists presumed to function without morphology, and the morphologists without genetics. One of the saddest sentences written by a Nobel Prize winner in Medicine or Biology was that by Hans Spemann[14] after he received the Nobel Prize for his marvelous discoveries on the organizer principle/region in the embryo when he said in the introduction of his book on the subject: "developmental physiology ... thus, the causal elucidation of developmental processes." Nowhere in the writings of this giant, or similarly minded workers in this field, e.g., Paul Weiss, does one find the word "gene," or a true causal, i.e., genetic, analysis of development.

It is of immense gratification to those of us who began work in morphology, as I did in Iowa City in the early 1950s in zoology and embryology (no course in genetics existed then for a major in zoology), that now all of the children of "mother" morphology, e.g., anatomy, embryology, cytology, paleontology, and genetics, are happily reunited through the means of molecular biology which has made it possible to study the expression of genes in various parts of the embryo while the developmental events are happening.

---

[13]W. Roux, *Programm und Forschungsmethoden.*

[14]H. Spemann, *Experimentelle Beiträge zu einer Theorie der Entwicklung* (Berlin: Springer, 1936).

Development

Development, or ontogeny, refers to the (biological) process of the coming-into-being of living organisms from earlier, undifferentiated stages, to embryonic, fetal, and immature stages, to the final attainment of sexual maturity, allowing the repetition of the haploid/diploid life cycle. Conventionally, the process in humans/mammals is divided into four stages.[15] Representative early stages of embryogenesis in various animals are illustrated in Figures 1–8.

I. *Pregenesis* (or progenesis, or pro-ontogenesis) refers to all of those stages from

- the establishment of the "*Keimbahn*," i.e., the separation of somatic and germ cells during earliest parental development, to
- the migration of the primordial germ cells to the primitive parental gonadal ridges,
- the differentiation of these ridges into testes or ovaries,
- the formation of egg and sperm cells during the process of meiosis, i.e., the genetic mechanism whereby the *diploid* number of chromosomes characteristic of the species (46,XX or 46,XY in humans) is reduced to the *haploid* number (23,X or 23,Y) in the functionally competent ovum or spermatozoon, to
- the process of fusion of male and female germ cells (fertilization), to
- the process of fusion of male (23,Y or 23,X) and female (23,X) pronuclei (karyogamy) leading to the reestablishment of diploidy and the first cell division initiating the process of development proper.

Thus, it is evident that the generations of living organisms are connected in an unbroken cycle of pregenesis-ontogenesis reaching back to the earliest ancestral forms of life on earth.

II. *Blastogenesis*[16] is the process extending from the first cell division to the end of gastrulation, in humans from day

---

[15]J.M. Opitz, "Blastogenesis and the 'primary field' in human development," *Birth Defects* OAS 29.1 (1993): 3–37.

[16]J.M. Opitz et al., "Defects of Blastogenesis" in *Human Development and Malformations: American Journal of Medical Genetics/Seminars in Medical Genetics*, ed. J.M. Opitz (American Journal of Medical Genetics, in press).

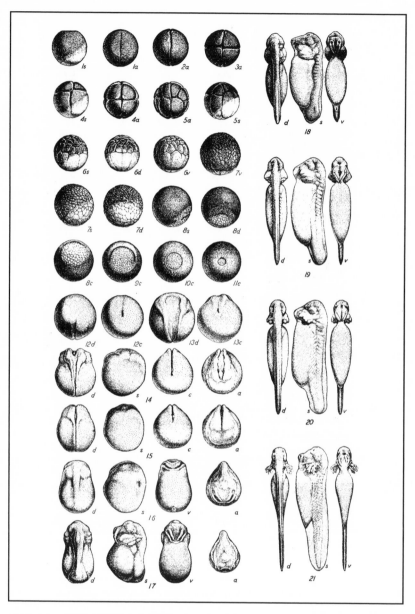

Figure 1. Development of the frog from first cell division (la) to the formation of the free-swimming larva (21). Also a marvelous example of the aesthetic standards and quality of art work of a substantial number of nineteenth and twentieth century morphologists/embryologists, in this case by my revered teacher Emil Witschi whose course in vertebrate embryology I took (1954) before Witschi's textbook *Development of Vertebrates* was published (Philadelphia: WB Saunders, 1956).

138

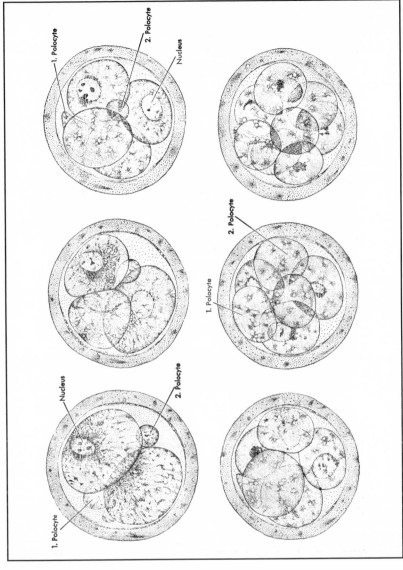

Figure 2. Early cell divisions (cleavages) of the rhesus monkey *(Macaca rhesus)* embryo initiating the process of blastogenesis, i.e., formation of the blastula or initial hollow sphere of embryologic development universal to all multicellular animals. Redrawn by Alfred Huettner from the work of Lewis and Hartman and published as Figure 159 in Huettner's textbook of embryology *(Fundamentals of Comparative Embryology of the Vertebrates*, rev. ed., 4th printing [New York: Macmillian, 1953]) which was assigned in Witschi's course in 1954, and which is one of the very few contemporary texts that contains an historical account of the concept of developmental fields.

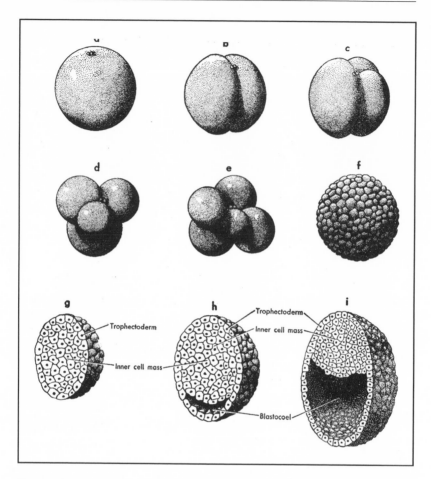

Figure 3. Huettner's schematic representation of early development in mammals, including cleavage and blastogenesis, the inner cell mass being the part of the embryo in which only very few cells give rise to the definitive embryo, all the other cells being involved in the ultimate establishment of the embryonic/fetal-maternal connection.

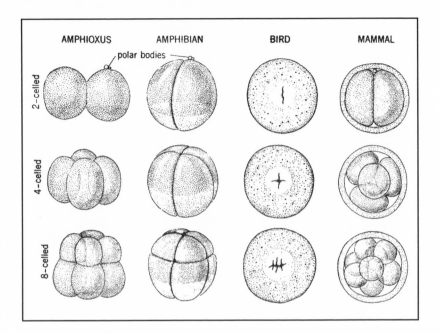

Figure 4. Comparing corresponding stages of development in *Amphioxus* (the lancet "fish," a protochordate), an amphibian (such as a frog), a bird, and a mammal, such as the rhesus monkey or human (N.J. Berrill and G. Karp, *Development* [New York: McGraw-Hill, 1976]). In birds cleavage is incomplete because of the huge amount of yolk. The next time you see a red spot on the yolk on one of your omelets eggs you might take a good magnifying glass to it to obtain a glimpse of the earliest stages of chick development.

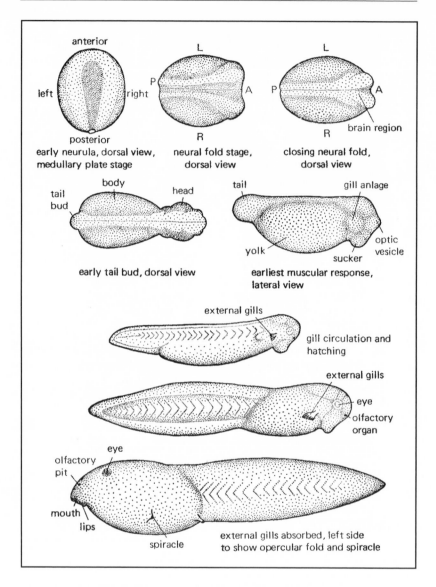

Figure 5. Neurulation and postneurula stages of development in the frog. In order to live on land, such an animal has to develop legs and lungs, and resorb gills and tail, i.e., has to undergo *metamorphosis*, a stage comparable to human metamorphosis from embryo to fetus at the beginning of fetal life. In humans most of prenatal life is fetal life (30–38 weeks), whereas in marsupials "fetal life" occurs in the "marsupium" or pouch (Berrill and Karp, *Development*).

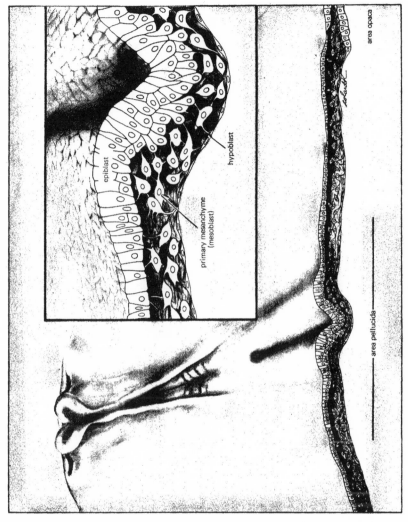

Figure 6. Gastrulation, in this case in the chick, a universal process in all multicellular animals occurring at the midline and leading to the definitive 3-layer embryo consisting of ectoderm, mesoderm, and endoderm (in this case still called epiblast, mesoblast, and hypoblast, respectively). The discovery of the universality of germ layer formation in vertebrates was one of the great triumphs of early nineteenth-century embryology beginning with Pander (1817) and von Baer (1828/1837). (Berrill and Karp, *Development*).

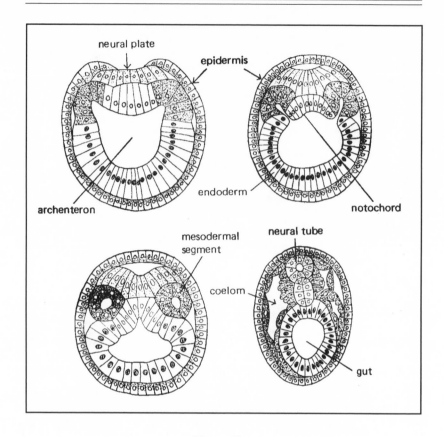

Figure 7a.

Figure 7a and 7b. Schematic representation of the formation of the initial embryonic parts in a "simple" protochordate, *Amphioxus*, the extraordinary aspect of those processes being that identical molecular "machinery" is involved in the early development of this ancient animal and in human beings, an impressive demonstration of the phenomenon of descent (Berrill and Karp, *Development*).

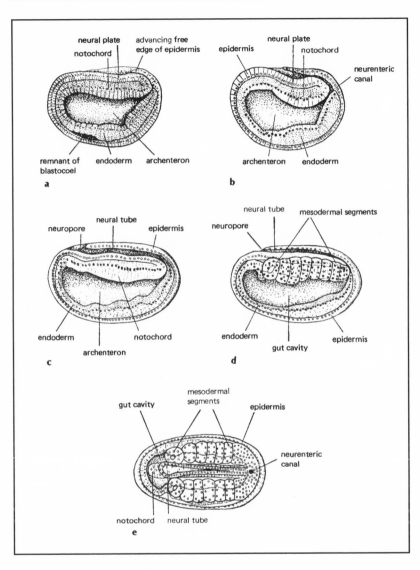

Figure 7b.

145

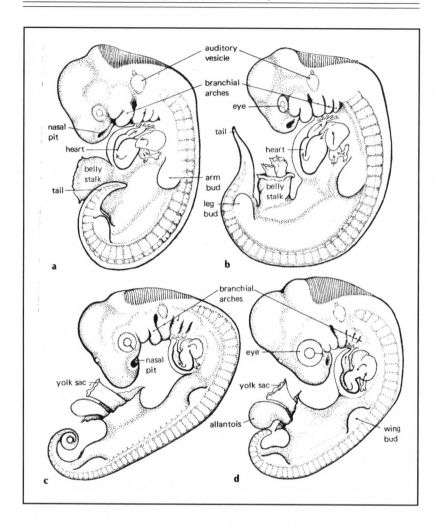

Figure 8. Impressive demonstration of homology of embryonic structure, hence of developmental mechanisms, based, as we now know, on the "deployment" of identical molecular inductive or pattern-forming systems, present by virtue of descent (with modification) from a common ancestor with prototypic developmental pattern. In this case, illustrating corresponding stages of development of the human (a), pig (b), reptile (c), and bird (d) (Berrill and Karp, *Development*).

one to day twenty-eight, stage one to stage thirteen, and including:

- the implantation of the early embryo into the wall of the maternal uterus and establishment of a maternal/embryonic/fetal blood circulation (Figure 9);

- establishment of the embryo proper, initially as a pluripotent, undifferentiated mass of cells (the inner cell mass) which early on acquires a definite polarity with three body axes (anterior-posterior/head-tail; dorsal-ventral/back-belly; right-left);

- establishment of *midline*, a morphogenetically active (not passive) embryonic landmark of crucial importance in early embryogenesis which gives rise to a dorsal midline structure called the *primitive streak* which, in turn initiates the process of *gastrulation* that leads to the formation of the three classic germ layers identified in the nineteenth century, namely, *ectoderm* (which gives rise to the nervous system, sensory organs, skin with its appendages and derivatives, and neural crest of vital importance in the formation of part of the heart, the cranio-facial skeleton, structures of neck, and the peripheral nervous system); *mesoderm* (which gives rise to the early heart, all connective tissues, renal/gonadal system, and axial and limb skeleton); and *endoderm* (which gives rise to lungs, gastrointestinal system, liver, spleen, etc.);

- the mesoderm on either side of the midline and incipient nervous system (the neural tube) segments into structures designated *somites* which ultimately give rise to segmented muscular, vertebral, and excretory (nephric/renal) organs. Face and neck are initially similarly segmented into structures called *branchial arches* homologous, at least during development, to the (embryonic) gill arches of our earliest fish ancestors. The arterial system arising from the heart initially is similarly segmented and bilaterally symmetrical, corresponding to the phylogenetically ancient arteries of each branchial (gill) arch;

- formation of the heart in an astonishingly precocious manner so that at seventeen days after fertilization and the initiation of embryogenesis, a contractile, i.e., beating tubular structure, is present![17]

[17]J.M. Opitz and E.B. Clark, "Heart Development: An Introduction," *American Journal of Medical Genetics: Seminars in Medical Genetics* 97.4 (Winter 2000): 238–247.

- formation of the central nervous system in the dorsal midline ectoderm, initially as a narrow midline plate (broader in the prospective head area) which then forms a rain-gutter-shaped structure, the edges moving or rolling up and toward the midline to fuse into a seamless, hollow (neural) tube extending from front to end of the embryo.

III. *Organogenesis* is the developmental process from day twenty-eight to day fifty-six after fertilization, from stage thirteen to the end of stage twenty-two, from the middle to the end of embryogenesis and the beginning of the fetal period and the process of *phenogenesis*. At that point, when the human embryo has a crown-rump length of three centimeters, all marsupial mammals (i.e., kangaroos, opossums, etc.) are born and complete fetal development in the mother's pouch, mouth firmly attached to a life-giving nipple. Organogenesis consists of two processes: the formation of organs and the other parts of the body in what is referred to as the secondary (or epimorphic) fields; and histogenesis (later differentiation of cells and tissues).

IV. *Phenogenesis* is the developmental stage from the transition of the embryonic to the fetal period (called metamorphosis), from the beginning of the ninth to the end of the thirty-eighth week after fertilization (fortieth week of "pregnancy" adding the two weeks from the time of fertilization to the time of the first missed period). Phenogenesis involves growth and final attainment of all of those qualitative and quantitative traits constituting family resemblance and racial affiliation.

V. *Growth and Development*: American Pediatrics speaks of its mission as "caring for the child during its period of growth and (psychomotor) development," i.e., from birth (however premature) to the age of puberty, i.e., the completion of sexual maturation to late adolescent stages. The time from birth till puberty really constitutes a continuation of phenogenesis with initial adaptation of the cardio-pulmonary system to extra-uterine life, and continued functional and histological maturation of many organs, including gonads, genitalia, and secondary sexual characteristics.

Developmental Fields

Developmentally, what connects us most essentially to our more or less remote ancestors? Here I will contend that it is the fact that morphogenesis (i.e., developmental processes) occurs in a *regionalized* manner, i.e., in parts or territories of the embryo which eventually give rise to final structure and in

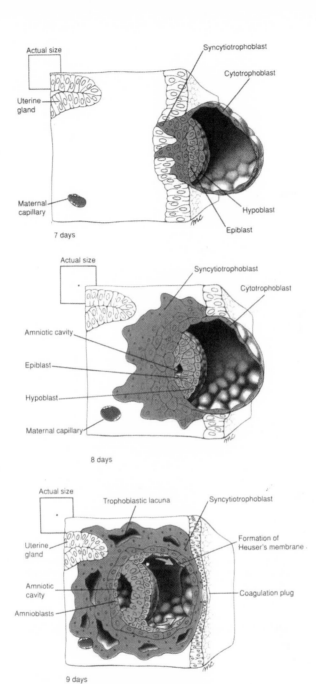

Figure 9. Implantation of the human ovum/blastocyst at seven days, and subsequent events on day eight and nine. Reprinted from *Human Embryology*, 2d ed., W.J. Larsen (New York: Churchill Livingstone), Copyright 1997, with permission from Elsevier Science.

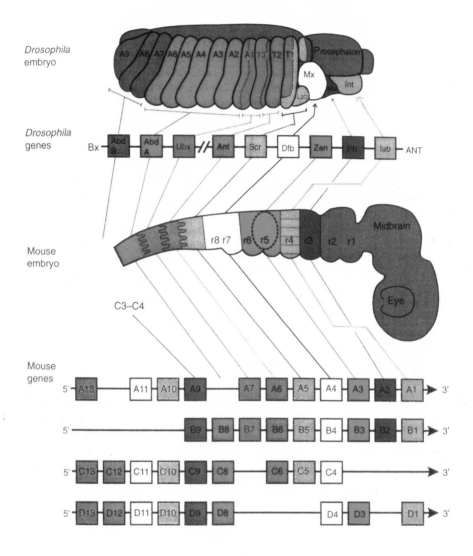

Figure 10. Dramatic illustration of the principle of universality showing identical molecular means (in this case the *Hox* genes) to organize the central nervous system in the mouse embryo and the anterior-posterior body plan in an invertebrate—here the fruit fly *Drosophila*. Note that between fly and mouse, there had been a quadruplication of the *Hox* gene cluster. Reprinted from *Human Embryology*, 2d ed., W.J. Larsen (New York: Churchill Livingstone), Copyright 1997, with permission from Elsevier Science.

which *processes are spatially coordinated, temporally synchronized, and epimorphically hierarchical.* During those glorious days of experimental embryology in the 1920s, these parts, or morpho-genetic units, came to be called *fields* by Alexander Gurwitsch, Paul Weiss, and Hans Spemann (who received the Nobel Prize in 1935 for this discovery), whose successors remained frus-trated until very recently in their attempts to elucidate the causal nature of overall and regionalized developmental pro-cesses.

Based on the work of these pioneers, one now speaks of three stages of field morphogenesis[18]:

- The *primary field* is the entire embryo before regionalization has begun. Some of the most important processes in the primary field, i.e., axis formation, are determined by the expression of maternal genes.

- The *progenitor fields* are the results of the initial regionalization of developmental processes in the early embryo before any regional parts (e.g., limbs) are grossly visible. They are the result of (are caused by) local expression domains of several major ("upstream") genes that control many subsequent complex developmental events encoded by a cascade of successively activated ("downstream") genes (and the silencing of others).

- *The final* or *secondary,* or *epimorphic* fields are the result of downstream gene activation/inhibition events, which progressively subdivide the progenitor fields into ever smaller embryonic units giving rise to the final structure. Thus, the (anterior) limb progenitor field, discovered in 1918 by the American embryologists Ross Granville Harrison and Samuel Detweiler, initially invisible in the shoulder/flank region of the embryo, represents the expression domain of several very ancient genes which cause the interaction between lateral plate mesoderm and overlying ectoderm to eventuate in the outgrowth of the limb bud which is eventually subdivided in a progressive proximo-distal, i.e., shoulder-to-fingertip manner into arm (A), A+ forearm (AF), AF and wrist (AFW), AFW and five metacarpal bones (ATWM), and finally into AFWM and

---

[18]M.L. Martínez-Frías, J. Frías, and J.M. Opitz, "Errors of mor-phogenesis and developmental field theory," *American Journal of Medi-cal Genetics* 76.4 (April 1, 1998): 291–296.

two phalanges in thumbs, three in each finger, all with their appropriate muscles, nerves, and blood vessels, and final dorsal emergence of a nail on each digit. A field remains a field so long as its developmental fate can still be (causally) modified by genes (mutant and non-mutant) or environmental causes (e.g., thalidomide).

Development and Evolution

It is an axiomatic "truth" (fact) that all morphogenesis occurs in developmental fields which may be discovered incipiently even in unicellular animals which may undergo regional morphogenetic specification into feeding, locomotory, and reproductive parts.[19] Evolution represents the outcome of permanent modification of prenatal developmental processes subsequently more or less fixed through selectively advantageous combinations of genes, not "new" genes, but novel combinations (or duplications) of pre-existing genes. *Thus, the developmental field is the fundamental unit of ontogeny and of phylogeny, i.e., of development and of evolution.*

During millennia of evolution, embryonic development became constrained into only a very few, final, and common morphogenetic (qualitative) pathways; thus, most of us have only one head, four limbs, thirty-four vertebrae, two eyes/lungs/kidneys, one liver, etc. Correspondingly, the repertoire of possible malformations is similarly limited, most having been known since antiquity. *Malformations* arise during blastogenesis and/or organogenesis. Most are anomalies of incomplete development, some of "abnormal" development. Anomalies of incomplete development were already recognized as such by Harvey[20] and represent persistence *(vestigia)* of embryonic states such as cleft of lip/palate, incomplete closure of neural tube (spina bifida, anencephaly), webbing of fingers/toes, persistence of the communications between the atrial or ventricular chambers of the heart, etc.

However, given the extraordinary antiquity of evolutionary developmental processes, their occasional recurrence as such in modern species should not be surprising. Such anomalies are called *atavisms* and were regarded as such already in

---

[19]G. Webster and B. Goodwin, *Form and Transformation* (Cambridge: Cambridge University Press, 1996), see especially part II "Fields and Forms," ch. 8 "The Unitary Morphogenetic Field," 193–230.

[20]G. Harvei, *Exercitationes.*

pre-Darwinian developmental biology (e.g., Lamarck's documentation of "baby teeth" in the Minke whale, a toothless baleen whale descended from a toothed, land-dwelling mammal ancestor; similarly, the rare occurrence of legs in whales or pythons). Six or seven fingers or even eight or nine toes in humans may impress as an instance of abnormal development, since humans normally have only five digits per hand/foot, until it is remembered that *none* of our earliest Devonian land dwelling ancestors (e.g., *Acanthostega, Ichthyostega, Tulerpeton)* had five digits, but rather six, seven or eight on fore- and/or hindlimbs. Cleft palate is abnormal in humans but normal in all birds and reptiles (except for crocodiles). Absence of corpus callosum (the large fiber bundle connecting right and left halves of the brain) is definitely abnormal in humans but normal in echidna and platypus, all marsupials and members of the rabbit family (lagomorphs); penoscrotal inversion, i.e., scrotum anterior to penis, is normal in marsupials and lagomorphs. In recent years it has been postulated several times that atavisms are recurrences of ancestral developmental states; indeed, the late Helen Taussig of Johns Hopkins Hospital, a pediatric cardiologist of Blalock-Taussig shunt fame, had asserted that all nonsyndromal malformations were what she called "ancestral anomalies."[21]

Everything that develops has evolved; nothing in development, whether normal or abnormal, can occur that evolution has not made possible. The ontogeny, structure, and function of every part of the body involves thousands of genes; hence, all malformations, whether mild or severe, are *causally heterogeneous* and nonspecific. The upstream molecular induction systems that establish the progenitor fields, say of heart, brain, or limbs, are highly conserved and used universally in all vertebrates, the *phenomenon of molecular universality* (Figure 10). Since in a gross, initial way, mouse, frog, chick, or fish use identical molecular means to put together a body plan, organ, or other part of the body, experimental animals, intelligently and compassionately used, are of enormous value in studying and understanding human development, and complement what we may infer causally or pathogenetically from human malformations. Indeed, mouse mutations are increasingly useful in guiding our understanding of abnormal human morphogenesis.

---

[21]J.M. Opitz and E.B. Clark, "Heart Development."

Until recently, *the emphasis in human genetics was the analysis of cause.* Cause frequently is an individually unique phenomenon, i.e., trisomy 21 in a (nonfamilial) occurrence of Down syndrome, *SOX9* mutation in an intersex boy with campomelic syndrome, etc. However, recently the focus has shifted dramatically to a study *of pathogenesis,* i.e., the developmental mechanisms responsible for the malformation. For it is still to a large extent the study of abnormal development that has led to a better understanding of normal development and its evolution, which has permitted a truly astonishing and inspiring perspective on the antiquity of life on earth and its morphogenetic processes. Indeed, it has been shown that many of the developmental events in invertebrates, such as the fruit fly *Drosophila*, are truly homologous to those of vertebrates, especially the early events (Figure 10). Some of these discoveries are truly astonishing. For example, axis and "head" formation in the polyp *Hydra* involves the same *WNT* signaling molecule required for axis and head formation in mammals.[22]

We are woefully ignorant about the evolution of meiosis, an absolute prerequisite for the formation of egg and sperm cells and the initiation of new life. However, what is known is that *defects of pregenesis are the most common cause of prenatal death in humans,* causing an approximate ninety percent mortality of prenatal humans, mostly during the earliest stages of development, leading to spontaneous abortion around or shortly after the first missed period. Some sixty percent of spontaneously aborted embryos have a gross *chromosome abnormality* (extra or lacking chromosome or chromosome piece, or extra set of chromosomes—69,XXY, 69,XXX or 69,XYY). Some six percent of stillborn babies (born after twenty weeks of gestation or at a weight of five hundred grams or more) have a chromosome defect, as do 0.6% of all liveborn infants, e.g., those with Down, Turner, Klinefelter, and other syndromes. In contrast, only 3–5% of all liveborn infants have a malformation, i.e., a defect of blastogenesis or organogenesis. Thus, the major burden of mortality in humans is *not* due to postnatal diseases, accidents, homicides, or suicides, but overwhelmingly due to prenatal causes, i.e., defects of pregenesis. It is perhaps ironic that we owe much of our developmental integrity to genes, which when mutated at a later stage in life, cause cancer.

---

[22]B. Hobmeyer et al., "WNT signalling molecules act in axis formation in the diploblastic metazoan *Hydra*," Nature 407.6802 (September 28, 2000): 186–189.

Colophon

Human and medical genetics have been the beneficiaries of immense advances in knowledge made to a substantial extent during the study of genetically abnormal humans.

Without question this knowledge will lead to a profound and lasting transformation of the practice of medicine without obviating in the slightest the combined input of clinical geneticists needed to provide analysis of phenotype, interpretation of molecular data to patients and families, teaching of medical and graduate students, house staff, colleagues, and support groups, performing clinical research, and providing clinical genetic services for an ever-increasing and demanding segment of humanity. Advances in biological knowledge will make it ever more possible for clinicians to study evolution at the bedside.

Medicine and medical genetics will flourish to the extent to which basic and clinical scientists collaborate in a mutually supportive manner and the administrators of academic programs acknowledge the essential role of clinical geneticists in building bridges between patients and researchers while teaching and providing services.

Such services will be needed as long as somatic and germinal mutations, meiotic nondisjunction, sporadic multifactorial traits, chromosomes breakage, stochastic defects of morphogenesis, and a huge prenatal death rate continue to afflict a long-suffering humanity. And while the (genetic) medicine of the future will become ever more molecular, it is to be hoped that it will simultaneously become ever more humane, ethical, and socially responsible, helping those afflicted redeem their suffering, bestow meaning on their lives, and grant a reasonable measure of hope.

FIDES ET RATIO:
THE COMPATIBILITY OF
SCIENCE AND RELIGION

WILLIAM A. WALLACE, O.P.

Almost a half century ago, in 1954, I taught my first course on the philosophy of nature. I have been doing so ever since, even though the courses are now called philosophy of science.[1] In the 1950's my teaching was based on the Fifth Lateran Council,[2] whose decrees had been reaffirmed by the First Vatican

---

[1]Philosophy of nature, as its name indicates, is concerned with the study of nature or of natural beings that exist outside the mind. Philosophy of science, on the other hand, is concerned with the methods and concepts of modern science, which are a type of knowledge and as such exist in the mind. The two are not to be identified, although there is considerable overlap in the materials covered in the two disciplines. For a study of the complementarity that exists between them, see W. A. Wallace, *The Modeling of Nature: Philosophy of Science and Philosophy of Nature in Synthesis* (Washington, D.C.: The Catholic University of America Press, 1996).

[2]The Fifth Lateran Council was an ecumenical council convened under Pope Julius II and continued under Pope Leo X. The materials

Council[3] and were reinforced by the encyclical *Humani generis* of Pope Pius XII.[4] All continued tranquil until the momentous years 1962–1965, the period of the Second Vatican Council.[5] As efforts were made to implement the decree *Optatam totius* of that council,[6] we found that drastic cuts had to be made in the philosophy program. Philosophy of nature, of course, became the first casualty. That was only the beginning, however, for soon, in seminaries throughout the country, more and more credits were taken away from philosophy.[7] Finally it was not unusual to find seminarians being admitted to theology programs without a single credit in philosophy.

---

that will interest us in this essay are contained in the Bull *Apostolici regiminis* of that council, issued in Session VIII, December 19, 1513, and concerned with the Church's teaching on the human soul, DS 1440–1441. See note 25 below.z

[3]The First Vatican Council was an ecumenical council convened under Pope Pius IX. The materials that will interest us come mainly from the dogmatic constitution *Dei Filius*, issued in Session III, April 24, 1870, and concerned with the Catholic faith, DS 3000–3045.

[4]Issued on August 12, 1950, and concerned with the crisis brought about by new tendencies in the sacred sciences, *AAS* 42 (1950), 561ff.

[5]The Second Vatican Council was an ecumenical council initiated by Pope John XXIII on May 17, 1959, and concluded under Pope Paul VI on December 8, 1965. The council itself consisted of four sessions or periods, the first extending from October 11 to December 8, 1962; the second, from September 29 to December 4, 1963; the third, from September 14 to November 21, 1964; and the fourth, from September 14 to December 8, 1965.

[6]This was the decree of the council on priestly formation. It specified that philosophy be taught to seminarians in such a way as to convey a solid and coherent understanding of man, of the world, and of God. It further recommended that its teaching be based on a philosophical heritage that is perennially valid, that it include contemporary philosophical investigations, especially those influential in the seminarian's own country, that it be abreast of recent scientific progress, and that it feature a critical study of the history of philosophy.

[7]The number of credits assigned to courses in philosophy in the United States varied throughout the country. Before the council, in orders devoted to the intellectual life such as the Dominicans and the Jesuits, preparation for the priesthood consisted of three years in philosophy and four years in theology, with an optional fifth year for pastoral training. After the council, the American Catholic Philo-

In such a situation you can imagine my joy at John Paul II's issuance, in 1998, of his encyclical *Fides et ratio* on the relationship between faith and reason.[8] Truly a magisterial document, this encyclical traces the importance of philosophical thought in the development of Christian doctrine. It depicts occasional periods of tension between faith and reason, but stresses the gradual realization within the Church of the complementarity that must exist between the two. This assumes that both be granted their autonomy, but, at the same time, it insists on their interdependence.[9] *Fides et ratio* points to St. Albert the Great and St. Thomas Aquinas as the first to stress this twofold relationship.[10] From their time onward, however, the encyclical also points out how philosophy has gone its separate way in the modern era, first proposing itself as a substitute for faith and now, in our own times, as its implacable enemy.[11] More important for our purposes, in several places *Fides et ratio* urges a prompt return to the teachings of St. Thomas Aquinas as an antidote to the ills of the present day.[12] These teachings are of particular importance for a correct understanding of the relationships that should exist between faith and reason, and thus between science and religion.

---

sophical Association recommended fifteen credits, or two years, for the teaching of philosophy in seminaries. A plan for covering the matter suggested by *Optatam totius* in two years, based on materials contained in the *New Catholic Encyclopedia*, will be found in W. A. Wallace, *The Elements of Philosophy: A Compendium for Philosophers and Theologians* (New York: Alba House, 1977), 9.

[8]The encyclical was issued by Pope John Paul II on September 14, 1998, the twentieth year of his pontificate. Citations herein are from the Vatican English translation as found in *Encyclical Letter Fides et Ratio of the Supreme Pontiff John Paul II to the Bishops of the Catholic Church on the Relationship between Faith and Reason* (Boston: Pauline Books and Media, 1998).

[9]See *Fides et ratio*, ch. IV, "The Relationship between Faith and Reason," nn. 36–48.

[10]Ibid., n. 45.

[11]Ibid., ch. IV, section entitled "The drama of the separation of faith and reason," nn. 45–48.

[12]Ibid., ch. IV, section entitled "The enduring originality of the thought of St. Thomas Aquinas," nn. 43–45. See also nn. 57, 58, 60 (note 84), 66 (note 89), 69 (note 93), 74, 78, and 82 (note 99). Note also, at the end of *Fides et ratio*, the reaffirmation of the teachings of Pope Leo XIII's Encyclical *Aeterni Patris*, 123.

The Compatibility of Science and Religion

In the modern mind, the juxtaposition of science and religion, seeing them as either opposed or as linked in some way, does not resonate significantly with Aquinas's thought. Paradoxically, much of what is now discussed under "science and religion" he would have seen as part of the larger problem of "faith and reason." Once the respective spheres of these two types of knowing are made clear, difficulties for him over the compatibility of science and religion quickly dissolve.[13]

In St. Thomas's view, faith means belief in God and acceptance of divine revelation as true.[14] Reason, on the other hand, refers to the way humans acquire knowledge by their natural powers of sense and intellect alone, without reliance on God or supernatural revelation. His distinction focuses more on the mode of acquisition of knowledge than on the knowledge acquired. A person whose reason is complemented by faith might thus be capable of knowing more truths than one who knows through reason unaided. But, if contradictory truths seem to derive from the two sources, then the competing claims of faith and reason have to be resolved, and one is faced with the typical controversy between science and religion.

To be more precise, faith for Aquinas is a supernatural virtue (along with charity and hope) that accompanies grace in the soul of the Christian and that disposes him or her to believe in truths revealed by God. Such truths are not self-evident to human reason, and assent to them must be determined by voluntary choice. If such a choice is made tentatively, it is called opinion. If it is made with certainty and without doubt, it is called faith. And since the object of faith is truth, which is the proper object of the intellect, St. Thomas saw faith as proximately an act of the intellect, even though it is prompted by an act of the will.[15]

Religion, like faith, is a virtue for Aquinas, but it resides not in the intellect but in the will.[16] It is allied to the virtue of

---

[13]This line of thought is worked out in some detail in W. A. Wallace, "Thomas Aquinas and Thomism," in *The History of Science and Religion in the Western Tradition*, ed. Gary B. Ferngren (New York and London: Garland Publishing, Inc., 2000), 137–140.

[14]A full discussion of faith by Thomas Aquinas is found in his *Summa theologiae*, II–II, Q. 1–7.

[15]Ibid., II–II, Q. 2.1.

[16]Ibid., II–II, Q. 81.

justice, which disposes a person to give to others their due. In the case of religion, the "other" is God. Those who observe their obligation to honor God as the author of their being are, in fact, religious persons. Being religious in this sense does not involve any special scientific knowledge and so does not bear on science vs. religion controversies.

Science is also a virtue for Aquinas, but it is a natural virtue of the human intellect.[17] It is a type of perfect knowing wherein one understands an object in terms of the causes that make it be what it is. It is attained by demonstration that meets the norms of Aristotle's *Posterior Analytics* and, as such, is certain and not revisable.[18] In no way dependent on divine faith, it falls completely outside the sphere of religious assent. But since most of what passes under the name of science in the present day is fallible and revisable, it would classify as opinion and not as science in Aquinas's sense.

Let us now apply these distinctions to three areas of conflict between science and religion and suggest how each has been or may be resolved. The first is the "Galileo Affair," relating to the alleged proof of Galileo Galilei that the earth moves; the second is the "Darwin Affair," relating to arguments of Charles Darwin for an evolution of species in the organic world; and the third to the problem of hominization, a problem still being worked on, whose solution bears on bioethical issues being confronted in the third millennium.

The Trial of Galileo

The dispute between Galileo and the Church over whether the earth moves or not is generally seen as the paradigm case of conflict between science and religion. The case has been extensively reviewed by historians, and most recently by a Galileo Commission appointed by Pope John Paul II in 1981, which presented its results to the Pontifical Academy of Sciences in 1992.[19] In brief, Galileo had made discoveries with the telescope that seemed to contradict statements of Scripture,

---

[17]Ibid., I–II, Q. 57.2.

[18]Aristotle, *Posterior Analytics*, I, 2; see also Aquinas, *Analytica posteriora*, I, 4–6, for his exposition of this teaching. For a clear explanation of the sense of the term "demonstration" in this context, see Melvin A. Glutz, C.P., "Demonstration," *New Catholic Encyclopedia*, vol. 4, 757–760.

[19]The discourse of Pope John Paul II to the Pontifical Academy of Sciences in presenting the findings of the Commission on October

particularly its teaching that the sun moves and the earth stands still. The Church immediately questioned his claim that the earth moves. Was this a scientific truth that could be experimentally demonstrated, or was it simply an hypothesis that accorded better with Galileo's observations? If a demonstration was possible, the Church said, he should convince other scientists of its truth; if not, he should leave the Scriptures to the Scripture scholars, for they are more expert than he in telling what the word of God means.[20]

As it turned out, by the time of his trial in 1633 Galileo was unable to offer direct proof of the earth's motion, and he was forced to recant and undergo the penalties imposed on him by the Inquisition.[21] Scholars are uniformly agreed that the penalties were unusually harsh and that Galileo suffered much at the hands of the Church. Still, the principle mentioned above for understanding the Church's position remains intact. Although theoretical arguments for the earth's motion were available in the seventeenth and eighteenth centuries, strict dem-

---

31, 1992, was published in the journal of the Pontifical Council for Dialogue with Non-Believers, *Athéisme et foi*, 27.4 (1992), 241–249. An English translation of the discourse appeared in the Weekly English Edition of *L'Osservatore Romano* of November 4, 1992. In an earlier discourse to the Pontifical Academy of Sciences on November 10, 1979, the pope stated that he wanted theologians, scholars, and historians to take a closer look at the Galileo case in order to openly recognize "wrongs from whatever side they come" and help dispel mistrust between science and faith (*AAS* 71 [1979], 1464–1465).

[20]The clearest adumbration of this teaching is contained in a letter written by Robert Cardinal Bellarmine to the Carmelite friar Paolo Foscarini (and Galileo Galilei) on April 12, 1615. An English translation of the letter is found in Maurice A. Finocchiaro, *The Galileo Affair: A Documentary History* (Berkeley and Los Angeles: University of California Press, 1989), 67–69.

[21]For an account of the trial and the bearing of demonstration on its outcome, see W. A. Wallace, "Galileo's Science and the Trial of 1633," *Wilson Quarterly* 7 (1983): 154–164. On Galileo's science, see W. A. Wallace, "Galileo's Concept of Science: Recent Manuscript Evidence," in *The Galileo Affair: A Meeting of Faith and Science*, eds. G.V. Coyne, M. Heller, and J. Zycinski (Vatican City: The Vatican Observatory, 1985), 15–35. This essay has been reprinted in *Studies in Philosophy and History of Philosophy*, vol. 15, *Reinterpreting Galileo*, ed. W.A. Wallace (Washington, D.C.: The Catholic University of America Press, 1986), 3–28, under the title "Reinterpreting Galileo on the Basis of His Latin Manuscripts."

onstrations of that motion were not available until almost a century later.[22]

One of the most important findings of the Galileo Commission, one that has generally been overlooked by scholars, is the circumstance that eventually led the Church to remove its prohibition against Copernicus, and along with that its prohibition against Galileo's teachings. This came about in 1820, when Pope Pius VII removed the prohibition at the request of Benedetto Olivieri, O.P., Commissary of the Holy Office.[23] Olivieri told the Pope that actual demonstrations of the earth's motion had by this time already been achieved, and so there was no reason to continue the prohibition. The demonstrations to which he pointed were of two kinds. The first was the work of a professor at Bologna, who, in experiments performed at Bologna between 1789 and 1792, dropped objects from a high tower and measured their deviation to the east, an indication of the earth's revolution on its axis. The second was the work of an astronomer at the Collegio Romano, who, in a work dedicated to Pope Pius VII and published in 1806, announced a measurable parallax for star Alpha in constellation Lyra. Thus he offered what Olivieri referred to as "a demonstration to the senses" (*una demonstrazione sensibile*) of the earth's annual motion around the sun. Significantly, both of these experiments antedated Friedrich Bessel's parallax measurement of 1838 and Léon Foucault's experiments with the pendulum of 1851, now most commonly cited as offering direct proofs of the earth's motion. Contrary to common opinion, the Church was thus in advance of its time in its use of scientific evidence.

---

[22]By "theoretical arguments" we mean proofs based on theories that had been partially confirmed and thus offer indirect evidence of the earth's motion. Some historians would cite Sir Isaac Newton's *Principia*, or, more fully, his *Mathematical Principles of Natural Philosophy*, first published in 1687, as offering such a proof, although Newton himself regarded the earth's motion simply as a hypothesis. Others would cite a measurement made by the English astronomer James Bradley in 1728 that detected the aberration of starlight as such a proof. More direct proofs were not found until much later, as discussed below.

[23]For details, see Walter Brandmüller and Johannes Greipl, *Copernico, Galilei, e la Chiesa: Fine della controversia (1820), gli atti del Sant'Ufficio* (Florence: Leo S. Olschki Editore, 1992). An abbreviated version of the proofs is given in W. A. Wallace, "Galileo's Trial and Proof of the Earth's Motion," *Catholic Dossier* 1.2 (1995): 7–13.

The Origin of Species

With regard to Darwin's *Origin of Species*, prior to 1859 all religious thinkers held that God created living species as they presently are found in the universe, fixed in species and immutable in essence. When Darwin published his controversial work, the Church in Rome did not take a public stance against him, possibly because of the lesson it had learned from the recently concluded Galileo Affair. Although Catholics were as opposed to Darwin's teaching as were Protestants, no official condemnation of the *Origin of Species* was ever issued from Rome. Over the next century, moreover, as more and more evidence accumulated in favor of evolution, the Church's attitude with regard to Darwin's thesis underwent a series of transformations.[24]

In 1870 the First Vatican Council addressed the problem of "faith and reason," but it was content to repeat the general teaching of the Fifth Lateran Council, namely, that the same God reveals the mysteries of faith and illuminates human reason with its truth, and one truth cannot contradict the other.[25] In 1909 the Pontifical Biblical Commission refused to call into question the literal and historical meaning of Genesis in cases "which touch the fundamental teachings of the Christian religion."[26] It ruled, however, that one is not bound to seek for scientific exactitude of every expression in the first chapter of Genesis, and that free discussion of the six days of creation is permitted.[27] These replies were further elaborated some forty years later in a letter of the Pontifical Biblical Commission to Cardinal Emanuel Suhard, archbishop of Paris, dated January 16, 1948. In it the Commission made clear that its earlier re-

---

[24]For brief overviews of the Church's developing position on evolution, see the following entries in the *New Catholic Encyclopedia*, 15 vols. (New York: McGraw-Hill, 1967), "Evolution, Human," vol. 5, 676–585; "Evolution, Organic," vol. 5, 685–694; and "Soul, Human, Origin of," vol. 13, 470–471; also the entry in the supplementary volume for the years 1967–1974, "Evolution (Some Philosophical Dimensions)," vol. 16, 175–177.

[25]Henricus Denziger and Adolfus Schönmetzer, eds., *Enchiridion symbolorum definitionum et declarationum de rebus fidei et morum* (Freiburg-im-Breisgau: Herder, 1965), 3017; abbreviated throughout as DS and the number.

[26]DS 3512–3514.

[27]DS 3518–3519.

plies were in no way a hindrance to further truly scientific examination of evolution "in accordance with the results acquired in these last forty years."[28]

Closely following on this letter, on August 12, 1950, Pope Pius XII issued the encyclical *Humani generis*, which elaborated more fully on what the Commission had in mind. It explicitly condemned materialism and pantheism, and counseled both caution and moderation in reinterpreting Scripture. But with regard to evolution the encyclical clearly recognized this as a valid hypothesis for explaining how the human body took its origin from previously existing living matter.[29] It further was not opposed to anthropologists and theologians engaging in additional research and discussion on the details of evolutionary teaching.[30] But the Pope pointed out two important qualifications that would place limitations on such discussions. The first was that they not call into question the Church's teaching that, however the human body may have taken its origin, the human soul is immediately created by God.[31] The second was that the present understanding of Catholic teaching on original sin, arising from the sin of Adam, renders untenable the theory of polygenism, or "many Adams."[32]

Another forty-six years had to go by before the Church advanced a fuller clarification of these qualifications. This was in an address of Pope John Paul II to the Pontifical Academy of Sciences on October 22, 1996, in response to the Academy's proposing "the origins of life and evolution" as the theme of its discussions.[33] The Pope returned explicitly to the statements of *Humani generis* as well as to the findings of the Galileo Commission which he had endorsed in 1992.[34] With regard to *Humani generis* he stated that, taking into account the state of scientific research at the time, as well as the requirements of theology, the encyclical considered evolution a serious hypothesis,

---

[28]DS 3862.

[29]DS 3895.

[30]DS 3896.

[31]DS 3896.

[32]DS 3897.

[33]The address was entitled "Truth Cannot Contradict Truth," a phrase taken from the Fifth Lateran Council (see DS 3017). Surprisingly, the address is not mentioned in *Fides et ratio*.

[34]"Truth Cannot Contradict Truth," n. 3.

worthy of investigation and in-depth study equal to that of the opposing hypothesis.[35] Almost half a century after the encyclical, the Pope went on, new knowledge has led to the recognition of evolution "as more than a hypothesis."[36] Now it has earned the epistemological status of a scientific theory. While not completely confirmed and demonstrated, evolution is now seen as a unified explanation that accounts for many facts and data, showing how they may be interpreted and interrelated. Thus it merits more serious consideration.

In addition to this bow towards the data of science, the Pope also stressed that there were philosophical issues that need to be addressed.[37] In effect, he stated, several theories of evolution are now in vogue, depending on how one evaluates the mechanisms they involve and the philosophies on which they are based. Some theories of evolution are materialist, others are reductionist, yet others are spiritualist. The key issue facing the Church is one of assessing "the true role of philosophy and, beyond it, of theology," in understanding the evolutionary process.[38] The Church's magisterium, he went on, has definite ideas about the conception and origin of man and his spiritual soul, and these must be taken into account in arriving at any true theory of evolution. The Pope concluded by enumerating the main points of Catholic teaching on these matters as contained in the constitution *Gaudium et spes* of the Second Vatican Council and the encyclical *Humani generis*.[39]

Some Thomistic Concepts

At this point I would like to propose a synthesis of these statements within the context of my own recent work relating philosophy of science to the philosophy of nature within the Thomistic tradition.[40] The papal pronouncements we have been considering so far discuss man's constitution in terms of the

---

[35]Ibid., n. 4.

[36]Ibid., n. 4.

[37]Ibid., n. 4.

[38]Ibid., n. 4 (at the end).

[39]Ibid., nn. 5–7.

[40]The earliest account of this work was in a report of my research as a Fellow at the Woodrow Wilson International Center for Scholars in Washington, D.C., in 1984, which appeared with the title "Nature as Animating: The Soul in the Human Sciences," *Thomist* 49 (1985): 612–648. A fuller development is found in my essay "Nature,

concepts of body and soul. Within the Church's tradition there is an older terminology that may help us gain a more philosophical understanding of the soul concept. This uses the pair "matter-form" to replace the pair "body-soul" and proposes that the soul is the substantial form of the human body. The Ecumenical Council of Vienne, on May 6, 1312, confirmed this teaching in a negative way when it condemned the view that the rational and intellective soul is not the substantial form of the human body as erroneous and inimical to Catholic truth.[41] This same teaching was then reaffirmed by the Fifth Lateran Council, on December 19, 1513. Further directed against Averroist teachings, the Lateran council condemned the views that the intellective soul is mortal, that it is one for all men, and that it does not exist *per se* and essentially as the form of the human body.[42]

The vocabulary of these councils is clearly Aristotelian and Thomistic. To understand this, let us inquire into the nature of the matter that is joined to the human soul as a form in the constitution of a human being. In the present day we have somewhat naive ideas about matter, although the concept is now being examined more closely by physicists in light of the quantum and relativity theories. Within the Aristotelian-Thomistic tradition, the substantial form that is the human soul is not united to ordinary matter, matter that falls under our senses. Rather it is united to a basic stuff that underlies ordinary matter. The Greeks called this stuff *prōtē hulē* and the Latins *materia prima*, both meaning "first matter."[43] I shall defer to Greek usage in what follows and speak of this as "protomatter." The concept is difficult to understand, but for our purposes it may be thought of as something like the potential energy of modern physics. Not itself visible or with tangible properties, it can yet be understood as present and underlying all physical processes.[44]

Human Nature, and Norms for Medical Ethics," in *Philosophy and Medicine*, vol. 34, *Catholic Perspectives on Medical Ethics: Foundational Issues*, eds. E.D. Pellegrino, J.P. Langan, and J. C. Harvey (Dordrecht-Boston-London: Kluwer Academic Publishers, 1989), 23–53. The fullest description to date is in the first five chapters of Wallace, *Modeling*, 1–194.

[41]DS 902.

[42]DS 1440.

[43]Wallace, *Modeling*, 8–9.

[44]Ibid., 9, 28, 53–63.

An important concept related to protomatter and substantial form is the concept of nature. For Aristotle and Aquinas a thing's nature is closely related to its essence, for nature is the thing's essence as this is a principle of operation.[45] In light of this definition, the two components of a thing's essence may be referred to as its nature: protomatter is its nature as merely potential, and form, or more precisely, its natural or substantial form, is its nature as actual and determining.[46] Now, when discussing physical theories, as we shall be doing here, it is preferable to speak of the natures of things rather than their essences. Essence is a more metaphysical concept; it connotes something absolute and eternal. Nature is more adapted to the natural sciences, where we are discussing the activities and operations of things found in the world of nature.[47] From such activities we judge their individual natures, and so can say this is gold, that is an oak tree, a squirrel, and so on. We can also classify natures, and so speak of inorganic natures, plant natures, animal natures, and human nature.[48]

All four of these types of natures are principles of stability in the order of nature, and so we refer to them as stable natures. In addition to such natures there is also the possibility of transient natures. St. Thomas referred to this type of being as an *ens viale*, a being "on the way" or becoming.[49] He

---

[45]Ibid., 1, note 1.

[46]Ibid., 28–29.

[47]See the discussion of the "form of cat" in *Modeling*, 288–291. Previous to Darwin, taxonomists regarded natural species as essences that were immutable, eternal, indivisible, and necessary. In the present day, most biologists would agree that real specific discontinuity exists in nature, but they would replace the pre-Darwinian concept of essence with the more dynamic concept of relatively stable population. Thus natural species are not fixed essences, but rather interbreeding populations that are isolated reproductively from other interbreeding populations. See R.J. Nogar, "Evolution, Organic" in the *New Catholic Encyclopedia*, vol. 5: 691–692.

[48]Each of these is discussed in detail in ch. 2, 3, and 5 of *Modeling*. Ch. 4, entitled "The Modeling of Mind," treats problems of knowledge found in animal and human natures to prepare for the transition from animal nature to human nature.

[49]Aquinas, *Summa contra gentiles*, II, 89. Here, Aquinas is discussing intermediate souls or substantial forms that function in the generation of higher animals and humans. Of these he states that "they are not complete in species but are on the way (*in via*) to a species, and thus they are not generated with a permanent status,

did not have extensive knowledge of such entities, using the expression to refer mainly to seeds, *semina.* But, among the great discoveries of modern science is the uncovering of an astounding number of transient entities in the physical universe. I refer to the world of elementary particles, most of which have a transitory existence. Practically all of them are radioactive, with a short half-life, and those that are not are charged particles, whose very charge makes their independent existence somewhat tentative. Do such elementary particles have natures? If they are real, and exist in the physical universe, it seems that we must grant them some status superior to that of an *ens rationis.* But the natures we shall attribute to them are not stable natures, like those of elements, compounds, plants, and animals. Rather they are transient natures, those of entities that enjoy only a fleeting existence but still are part of the world of nature.

A Creation-Evolution Schematic

With this we are in a position to assess the epistemological status of evolutionary theory from the viewpoint of a renewed Thomistic philosophy of nature. To do so I shall make use of a diagram which I label "A Creation-Evolution Schematic," shown in Figure 1.[50] The diagram sketches a concordist view of Catholic theology and modern science, associating God's creative act at the beginning of time with the "big bang" theory of cosmic origins. Time t_o began some ten to fifteen billion years ago with the production by God, *ex nihilo*, of the primordial mass-energy of which the universe is now composed. Along with the act of creation, God as First Agent or Prime Mover also initiated the "big bang," releasing the enormous energy of the primitive mass for the formation of the natures now found in the uni-

---

but only that through them, the ultimate species may be arrived at." See W.A. Wallace, "St. Thomas on the Beginning and Ending of Human Life," *Studi Tomistici,* vol. 58, *Thomas Aquinas: Doctor humanitatis hodiernae* (Rome: Società Internazionale Tommaso d'Aquino, 1995), 394–407, especially at 396, note 2; also W.A. Wallace, "Nature, Human Nature, and Norms for Medical Ethics," 35–36, 42–44.

[50]An earlier version of this diagram is given in "Nature, Human Nature, and Norms for Medical Ethics," at 37, and explained on the following pages. There natures are indicated by the letters SF for substantial form. In subsequent drawings the letters NF, for natural form, have been used. Here we use simply "nature" as capturing the same idea.

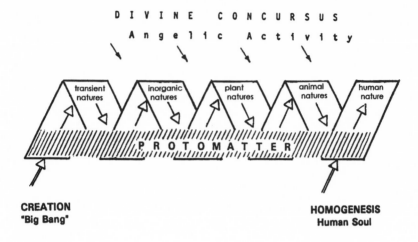

Figure 1. A Creation-Evolution Schematic

verse. This was a cosmic explosion, whose vestiges are still discernible at the edge of our expanding universe.

Cosmic evolution is slightly more difficult to explain. In the figure I give the elements of a generic picture by indicating in succession along the time axis the five types of natures just discussed. These are, in order, transient natures, inorganic natures, plant natures, animal natures, and human nature. They correspond to the stages of evolution commonly accepted among scientists: the period of fundamental particles impelled at high energy; that of element and compound formation; the two periods of biogenesis, wherein first plants and then animals are generated; and finally that of hominization, when *homo sapiens* first appeared.

The natures indicated in the upper parts of the parallelograms are actually substantial or natural forms (natures as actual); the lower parts of the parallelograms indicate their companion principles, protomatter, the basic stuff of the universe (natures as potential). Leaving aside, for the moment, the human form or soul, on the assumption that this is produced directly by God, as shown at the right of the diagram, we may ask where other natural forms come from. The answer supplied by St. Thomas is a surprising one: they are not preexistent as forms, nor are they created in any way; instead, they are simply "educed" or led forth from the potency of protomatter. This is his doctrine of the eduction of form from the potency of matter (*eductio formae ex potentia materiae*), which holds that all natural forms (man's excluded) are already precontained in the potentialities of the substrate, and require only the action of an appropriate agent to bring them forth into being.[51]

The analogy of the sculptor may help us understand this. We may ask where the form of David existed before it was carved out of marble by Michelangelo. One answer would focus on the exemplary cause: the form existed in the mind of the sculptor. But an equally valid answer would look to the material cause, to the block of marble, and say that David's form was resting in there all along, simply waiting to be led forth, educed, liberated from the matter under the action of Michelangelo's chisel. In a proportionate way, natural forms may be said to be resident in protomatter, their exemplars already present in the Divine Mind, awaiting only the proper agent to confer on them actual existence.

Now let us run through the various stages of development shown in Figure 1. At the initial creation, God created the transient natures we know as elementary particles. These we can still "recreate" in the present day by educing them from protomatter with high energy accelerators. From such particles were formed vast numbers of nuclear particles, from which the elements of the Periodic Table were generated—inorganic natures, the first stable natures in the universe. From these elements compounds were formed, and then from these stars and planets. That takes us up to the first five billion years of the universe's existence.

None of the development in our first two stages may be called "evolution" in the proper sense. Most of the changes can be duplicated by scientists in their laboratories, and the mechanisms that are involved are all well understood. It is in the last three stages that Darwin's *Origin of Species* comes to be involved. And yet some of this can also be explained on Thomistic principles. For St. Thomas, both plant and animal forms are what he would call "material forms," that is, forms that can be educed from the potency of matter. The genesis of living things from nonliving matter was not a problem for him or for other

---

[51]Aquinas uses this expression in his *Summa theologiae*, I, Q. 90.2, and in his opusculum *De spiritualibus creaturis*, when juxtaposing the way in which the human soul is produced directly by God, through an act of creation, to the way in which other forms are educed from matter under the causal action of appropriate agents. For a fuller explanation of this natural process, see Wallace, *Modeling*, 58–61.

medievals, since they took spontaneous generation as a fact and thought that it regularly took place in nature.[52] The difficulty lay not on the part of the matter, but rather on the part of the efficient agents, the cosmic forces that would be adequate to bring forth the higher forms of the living from inanimate matter. Here St. Thomas's belief that the angels took part in the governance of the universe provided an adequate answer for him.[53] And this surely would be an elegant role for these immaterial substances, namely, as agents that bring the lower natural forms of the universe to the summit of hominoid perfection, the stage immediately below that of the human race.

The final stage of cosmic evolution, then, is hominization, the appearance of man with his special type of natural form, an immaterial and immortal soul. Here there is a break in the line of causality extending back to creation, because the human soul, as immaterial and spiritual, cannot be educed from the potency of matter. The entire process of development and evolution, as diagramed thus far, can bring organisms to a level just below that of thought and volition, but they cannot progress to the final stage. Here God's creative act is required. So we indicate this with a second input of divine causality, the production, *ex nihilo*, of the human souls of our first parents, tailored to match the ultimate disposition of matter, as this has been prepared, over billions of years, for their reception.

The foregoing account, it should be stressed, gives only a generic picture of the evolution of natures, grouping all species of elements and compounds under inorganic natures, all kinds of vegetative life under plant natures, and all types of sensitive life under animal natures. Notice that we have not been talking about species as essences that are fixed and eternal, as older classifications regarded them. The modern biological con-

---

[52]See the appendices to my translation of questions 65 to 74 of the First Part of the *Summa theologiae*, Volume 10 of the 60-volume Gilby translation of the *Summa* entitled *Cosmogony* (New York: McGraw-Hill and London: Eyre & Spottiswoode, 1967), especially Appendix 6, par. 2, 198.

[53]*Summa theologiae*, I, Q. 110. For a discussion of Aquinas's position on this matter in the context of other medieval views, see James A. Weisheipl, O.P., "The Celestial Movers in Medieval Physics," *The Dignity of Science: Studies in the Philosophy of Science* (Washington, D.C.: The Thomist Press, 1961), 150–190, especially 183–190. A fuller account will be found in Thomas Litt, O.C.S.O., *Les Corps célestes dans l'univers de saint Thomas d'Aquin* (Louvain: Publications Universitaire, 1963).

cept applies the term species to interbreeding populations that are isolated reproductively from other interbreeding populations. This fits in well with the concept of nature, which admits of a developmental orientation such as is implied in an evolution of species. Organisms are not immutable, necessary, and eternal, as one might think of essences. Natures are principles of operation, and these are sufficiently changeable to ground a theory of evolution that allows for transformations of natural kinds.

How many natural species are there? A number that is infinite, potentially infinite, just as is the potency of protomatter. All species are latent within protomatter, but they become known to us only when conditions are appropriate for educing them from this material substrate. We discover them as we study the world of nature, and so we determine them empirically, both as they exist in the present day and as they may have existed in ages past. But there is no way of knowing their number *a priori*, short of our being able to read the mind of God.

The Problem of Hominization

This may perhaps suffice for our discussion of the theory of evolution. Portions of it, particularly those relating to the physical sciences, have been very well confirmed. Other parts are quite incomplete, since the mechanisms that have been invoked are far from being sufficient to explain the complex organisms that now populate the world of nature.[54] But as a working theory it has been very fruitful in areas of biochemistry, particularly those relating to genetics and the study of genomes of different species, plant, animal, and human. The human genome project, in particular, is now posing a series of problems that have to be addressed in the areas of genetic engineering, cloning, and similar procedures.[55] As background, let us sketch briefly St. Thomas's views on hominization, that is, how human beings are procreated by their parents through a process of human generation.

---

[54]For a discussion of these limitations, see Michael J. Behe, *Darwin's Black Box: The Biochemical Challenge to Evolution* (New York: The Free Press, 1996).

[55]A comprehensive introduction to the genome for the general reader is Matt Ridley, *Genome: The Autobiography of a Species in 23 Chapters* (New York and London: Harper Collins, 2000). A special report entitled "The Business of the Human Genome" will be found in the July 2000 issue of *Scientific American*, 48–69. Two studies that

Aquinas is fairly well known for his teaching that the beginning of human life is a gradual process, that the human soul is not infused into the incipient organism at fertilization but rather is prepared for by a succession of forms that dispose the matter for the reception of an intellective soul.[56] Less well known is his speculation that the reverse process may occur at the ending of human life, namely, that the human soul may depart from the body before all signs of life have disappeared from it.[57] Both views are opposed to what is commonly thought in Catholic circles, namely, that human life begins at fertilization, when the rational soul is infused by God into the body, and terminates at death, when the same human soul departs from the body.

With regard to human generation, Aquinas followed Aristotle in holding that the conception of a male child was not

---

address the genome from a Catholic perspective are: *The Human Genome Project: Proceedings of the ITEST Workshop, October 1992*, eds. Marianne Postiglione, R.S.M., and Robert Brungs, S.J. (St. Louis: ITEST Faith/Science Press, 1993), and *The Genome: Plant, Animal, Human: Loyola University, Chicago, August 1–5, 1999*, eds. Robert Brungs, S.J., and Marianne Postiglione, R.S.M. (St. Louis: ITEST Faith/Science Press, 1999). In a presentation at the Loyola University workshop entitled "Biogenetics and Technology: How Will the Race Be Won?," 92, Kevin FitzGerald, S.J., references an article in the April 1999 issue of *Scientific American* that is devoted to the replacement and repair of tissues in the human body. The article is part of a Special Report in that journal entitled "The Promise of Tissue Engineering," 59–89, and is entitled "Embryonic Stem Cells for Medicine," by Roger A. Pedersen, 68–73. Pedersen gives his personal views on "Ethics and Embryonic Cells" at 71 of that article.

[56]Aquinas goes into some detail on the process of human generation in four places: *In II Sent.*, 18, 2, 1–3; *Summa contra gentiles*, II, 86–89; *De potentia*, 3, 9–12; and *Summa theologiae*, I, Q. 118. All of these texts are analyzed by Michael A. Taylor in his "Human Generation in the Thought of Thomas Aquinas: A Case Study on the Role of Biological Fact in Theological Science" (S.T.D. diss., The Catholic University of America, 1981). A different analysis that relates Thomas's teaching to studies in modern embryology is Norman M. Ford, S.D.B., *When Did I Begin? Conception of the Human Individual in History, Philosophy and Science* (Cambridge: Cambridge University Press, 1989), 19–63.

[57]See Wallace, "St. Thomas on the Beginning and Ending of Human Life," 397, 403–407; see also W.A. Wallace, "Aquinas's Legacy on Individuation, Cogitation, and Hominization," in *Studies in Philosophy and History of Philosophy*, vol. 28, *Thomas Aquinas and His Legacy*, ed. David Gallagher (Washington, D.C.: The Catholic University of America Press, 1994) 173–193, esp. 188–193.

completed until the fortieth day after intercourse, whereas that of the female child was not completed until the ninetieth day.[58] Animation, for Aristotle, was immediate in the sense that a soul of some type was present as soon as the male's semen fertilized the material provided by the female, but this was not a human soul at the outset. In its earliest stage it was a nutritive soul (*anima vegetativa*), which regulated the early growth of the embryo. When development was sufficient to support sensation, the nutritive soul was replaced by a sense soul (*anima sensitiva*), and this in turn was ultimately replaced by the human soul. Aristotle recognized that it was difficult to determine precisely when the human soul came to be present in the embryo and simply stated that it came from outside and was divine.[59] St. Thomas picked up on this teaching and held that the intellective soul (*anima intellectiva*) was created by God and infused into the embryo at the completion of the developmental process, and that this soul once present performed all the functions of previous forms in the incipient organism.[60]

A notable feature of Aristotle's teaching on the earlier stages of animal generation is the role he assigned to the nutritive soul and the sense soul in the developmental process. In the case of the nutritive soul, he assumed that the semen and the unfertilized matter of the female, while still separated from each other, already possessed a nutritive soul, although they did so only potentially. Such an embryo thus lived the life of a plant, first with the parent drawing nourishment to it and then with the embryo beginning to nourish itself as a whole. When this second stage occurred the nutritive soul lost its potential status and came to be present actually.[61] Similarly, at the onset of animal life the sense soul was present only potentially. For it to become actually present sense organs had to develop in the organism, and particularly the sense of touch, so that the embryo could experience sensation. Since there can be various degrees of sensation, moreover, and in the higher animals all of the sense organs have to be developed before a specific animal is produced, Aristotle held that the developing animal embryo first became an animal in general and then one of a particular type, depending on the organs developed. The tran-

---

[58]Aristotle, *On the Generation of Animals*, I, 729a–744.

[59]Ibid., 736b.

[60]Aquinas, *Summa theologiae*, I, Q. 118.2, reply 2.

[61]Aristotle, *On the Generation of Animals*, 736b.

sition process would thus occupy a considerable period of time. That served to explain why the sensitive soul in the developing human being, the animal with the most refined sense powers, required forty days to reach its complete actuality.

St. Thomas apparently subscribed to this aspect of Aristotle's teaching and it is here that he introduced the concept of a transient entity, or *ens in via,* into the discussion. He did so when discussing the intermediate souls or substantial forms that function in the generation of higher animals and humans.[62] The successive replacement of forms in these cases takes place by a series of natural generations and corruptions. Thus, when leading up to the human soul, the form which is most perfect in nature, many intermediate forms and generations will be found. These intermediaries, Thomas writes, are not complete in species but are "on the way," *in via* to a determinate species.[63] So they are not generated with a permanent status but only transiently, so that, through them, the ultimate species may be arrived at.

Figure 2 now sums up this discussion, putting it in the same context we have employed in Figure 1. Essentially a similar diagram is involved along the top, showing the passage from the stable human nature of the parents, to a potential plant form, then to the plant form itself, then to a generic animal form that succeeds it, and finally the human nature of the offspring, a rational soul created individually by God, whose action is shown by the double-shafted arrow at the lower right. What is most remarkable, when one compares Figure 2 with Figure 1, is the way in which the nine-month generation of the individual human being seems to recapitulate the four billion year evolutionary process whereby the human species evolved from the most primitive life forms on our planet.

Delayed Hominization

The process explained by St. Thomas is known today by the expression "delayed hominization." It is opposed to the ex-

---

[62]Aquinas, *Summa contra gentiles*, II, 89.

[63]The Latin reads as follows: "Nec est inconveniens si aliquid intermediorum generatur et statim postdum interrumpitur: quia intermedia non habent speciem completam, sed sunt in via ad speciem; and ideo non generantur ur permaneant, sed ut per ea ad ultimum generatum perveniatur."

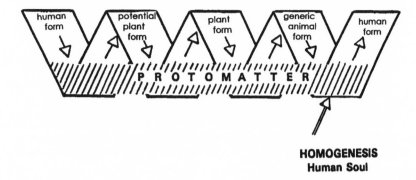

HOMOGENESIS
Human Soul

Figure 2. St. Thomas's View of Human Generation

pression "immediate hominization," which sees God creating the human soul and infusing it into the zygote at the instant when fertilization occurs. It is of interest to note that St. Thomas knew of immediate hominization, but did not see it as taking place in the generation of ordinary human beings. Rather, he saw it as taking place only in the Incarnation, that is, in explaining how the Second Person of the Trinity assumed human flesh and so became incarnate. Whereas ordinary people come into being by delayed hominization, for Christ's conception St. Thomas maintained that this was immediate and so that Jesus was conceived in miraculous fashion. His reasoning proceeded as follows.[64]

For Aquinas, the conception of Christ may be considered as taking place in three stages. The first is the movement of the blood to the place of generation; the second, the formation of the body from the matter available; and the third, its growth to perfection of size. The first, St. Thomas argues, could not be in an instant, for this goes against the nature of motion or movement, the parts of which come into place successively. So too, the third must be successive, because growth always entails motion and because it issues from a power of a soul that is acting in a body that is already formed and must needs operate in time. But the second stage, Aquinas continues, that of the actual formation of the body, in which the notion of con-

---

[64]Aquinas, *Summa theologiae*, III, Q. 33.1; see also *In III Sententiarum*, dist. 3, q. 5, a. 1.

(a) Fr. Ford's Theory Based on Individuation (14-day delay)

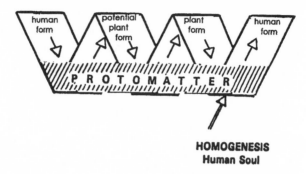

HOMOGENESIS
Human Soul

(b) Fr. Donceel's Theory Based on Cognitive Organs (several weeks delay)

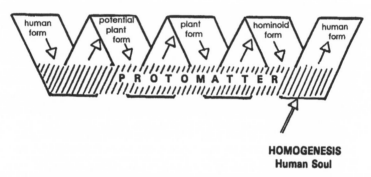

HOMOGENESIS
Human Soul

Figure 3. Recent Theories of Delayed Animation

ception consists, took place in an instant. This is so because of the infinite power of the agent, namely, the Holy Spirit, through which the body of Christ was formed. The greater the power of the agent the more speedily it can dispose its matter to the reception of a form. So an agent of infinite power can instantly fashion protomatter to its most perfect form, in this case, that of human nature. That is the sense in which Our Lord's conception was "immediate." It did not take place in time, as does that of ordinary human beings, but rather took place in an instant, through the all-powerful influence of the Holy Spirit.

What is important about this line of reasoning is that, for St. Thomas, even the miraculous conception of Christ had to respect a natural process, in this case, the need of protomatter to be previously disposed by an agent cause. Only when matter was properly disposed could it be united to Christ's human soul and the Incarnation brought to completion. In the order of nature protomatter cannot receive a natural form instantaneously. One or more agents must act on it, alter its dispositions, so that it becomes a fit counterpart for the form it is to receive. And, like all natural processes, such alteration requires time. Only in the case of Christ's conception was the need for such a temporal process set aside. And that was because of the miraculous action of the Holy Spirit on the matter in Mary's womb, which caused the human soul of Our Lord to be joined to protomatter in an instant.

To return, finally, to Aquinas's treatment of human generation, all of this was worked out on the basis of Aristotle's teaching as developed by medieval commentators, and particularly Avicenna.[65] Little empirical evidence was available to support the various steps of the argument. In our own day, however, the human reproductive process is being studied intensively, and there is an abundance of evidence that can be brought to bear on the problem of hominization.

Two lines of argument have been advanced that generally favor Aquinas's solution, now shown in Figure 3. The first, based on the possibility of twinning in the formation of the fetus, is essentially an argument from individuation.[66] This would pro-

---

[65]Details of Avicenna's teaching will be found in Taylor, "Human Generation."

[66]This line of argument has been developed in detail by Norman Ford in his *When Did I Begin?*, ch. 3 through 6. This book was printed in 1988 and reprinted in 1989 with minor corrections. In the 1989 edition Ford concludes: "Instead of viewing development in the first two weeks after fertilization as *development of* the human individual, I have argued the process ought to be regarded as one of *synthesis of* a human individual. We have seen that about fourteen days after fertilization, until the appearance of the primitive streak, the multiplying cells are naturally synthesizing a human individual. They have been aptly described as *personne en devenir*. The power of this incipient microscopic human individual to develop and grow from a tiny beginning to adulthood is paralleled by the adult person's ability to trace back one's personal history to the same beginning. A human individual and one's personal history begin together when a living ontological individual with a truly human nature commences development while ever remaining the same individual being.... If the thesis I have defended does emerge as the truth, I would suggest that the

pose that the definitive individuation of the human fetus does not occur until fourteen days after conception, and thus that the intellective soul, and so the human person, need not be present before that time. The second argument is based on the organ systems required first for sensitive life and then for the exercise of reason, which would involve the senses, the nervous system, the brain, and especially the cortex.[67] The time when such organ systems are present in the human fetus, especially the organization of the brain, must be ascertained by embryology. This probably occurs somewhere between several weeks and the end of the third month after conception, and so it is possible, on this theory, that human animation does not

---

term 'embryo' be applied from the primitive streak onwards. Prior to this I would suggest that the developing embryonic cells be referred to as 'proembryo' rather than 'preembryo' to indicate that though they have not yet become an embryonic human individual they are definitely developing towards that goal" (at 181–182, original emphasis). More recently, when arguments were being developed in the U.K. on the use of human embryos for stem cell research, Fr. Ford published a brief article in the *Tablet* of December 9, 2000 reaffirming his position that "the genetic code ... produces an organized living human individual about fourteen days after fertilization" (at 1672), but arguing against any use of human embryos for scientific research. It is of interest to note that the *Washington Times* of January 23, 2001 reported that Britain had legalized the cloning of human embryos, but stipulated that "the clones created under the new regulations would have to be destroyed after fourteen days" (at 1).

[67]This line of argument has been proposed by Joseph F. Donceel, S.J., in his "Immediate Animation and Delayed Hominization," *Theological Studies*, 31 (1970): 76–105. (For a survey of related opinions, see Gabriel Pastrana, O.P., "Personhood and the Beginning of Human Life," *Thomist*, 41 [1977]: 247–294.) Donceel bases his argument directly on St. Thomas's theory of delayed hominization. Thus he writes: "If form and matter are strictly complementary, as hylomorphism holds, there can be an actual human soul only in a body endowed with the organs required for the spiritual activities of man. We know that the brain, and especially the cortex, are the main organs of those highest sense activities without which no spiritual activity is possible" (ibid., 83, cited by Ford, at 51). Donceel then goes on: "The least we may ask before admitting the presence of a human soul is the availability of these organs: the senses, the nervous system, the brain, and especially the cortex. Since these organs are not ready during early pregnancy, I feel certain that there is no human person until several weeks have elapsed" (ibid., 101, cited by Ford, at 52).

occur before this time.[68] The alternatives are shown graphically in Figure 3.[69]

Both of these conclusions, if accepted, would have far reaching implications for future work in human genetics. It is not my intention here, however, to speculate on the validity of Aquinas's views in this important matter. Just as in the two previous cases of the earth's motion and the evolution of species, scientific evidence has already been brought to bear on an ultimate solution, so in the case of hominization, it seems that science will have much to say on how that problem should be resolved. Here is where the ultimate compatibility of science and religion can be tested. It is my hope that not only scientists, but philosophers of nature trained in the Thomistic tradition, will be on hand to participate in its solution.

---

[68]Ford's evaluation of Donceel's thesis reads as follows: "One weakness in Donceel's position is the unjustified demand for the formation of sense organs and of the brain for rational ensoulment once it is admitted there are no rational functions performed for at least two years. Insufficient reasons seem to be given to justify delaying rational ensoulment after conception and the formation of the individual embryo for some vaguely specified 'several weeks.' He might be right, but his philosophical arguments need to be supported by more solid embryological evidence to determine the minimum period of time after fertilization before which rational ensoulment or hominization could not possibly take place." (ibid., 52).

[69]For a discussion of immediate hominization see Mark F. Johnson, "The Moral Status of Embryonic Human Life and Moral Issues," in this volume, 181–198.

The Moral Status of Embryonic Human Life

Mark F. Johnson

The importance of this topic is matched by its complexity. On the one hand, the so-called "moral status of embryonic human life" is just another way of raising the question whether we are concerning ourselves with a living member of the human species, that is, whether or not we are concerning ourselves with a "person" in the customary sense of the word "person." If we are dealing with a person, then, when we fellow-human beings are at our best, that person will be granted all of his or her rights to life, equal treatment under the law, and so on. And of course, from our Catholic theological perspective, that also means a providentially governed destiny by God, and a vocation to holiness and perfection in the Christian community. It is hard to imagine a more basic, more "rock-bottom," foundation for rights and duties than the basic status of being human. And history teaches us what mischief ensues when people in power define away the humanity, the "personhood," of those who have little power, and little voice.

On the other hand, given the realist tradition of philosophy that has guided our Catholic thinking, the determination of whether the human embryo is a human being will be first and foremost an issue of observation, analysis, and judgment. And just as in other areas of our life we avoid prejudgments of fact, based not upon the evidence of things themselves, but rather upon our oftentimes legitimate worries, so also here, in this most important of matters, we Catholics must hold our justified concerns in check so as to avoid reckless prejudgment—which inevitably harms our credibility on this and other issues. Rather, we need to examine the facts of the human embryo offered to us by experts of observation, analyze these facts in accordance with the principles of faith and reason we have received from our forbears and from our own direct contact with the world. Then, finally we need to arrive at a judgment about whether the embryo is, simply put, a living member of our human family. All this information will become a foundational element in the moralist's consideration of the many topics with which we are concerned here in this workshop: abortion, rape protocols, stem cell research, and so on.

But perhaps it is not as easy as all that. In other words, perhaps it is not as easy as the biologist and philosopher studying the embryo, saying "yes, it's a human person," and handing the issue over to the moralist. For the disciplines that study the development of the embryo, and the disciplines that study the moral conduct of humans, both have in common the inherent difficulty of arriving at certitude in a domain of study whose subject is something characterized most by change. Change is the antithesis of knowledge. It was for this reason that the philosopher Plato abandoned the physical world to study the world of mathematics and his "Ideas," a world which he came to consider "the really real world." We Catholics, however, must take a different road. We were created as physical beings by God, who placed us in this physical world, and it is this physical world that our senses and intellect are best structured to understand, even if we should, through contemplation and prayer, on occasion catch a glimpse of what awaits us in heaven. So we are destined not only to walk in this valley of tears, but also to study it so as to learn from it. And what we do learn from it might well not attain the level of certitude that a Plato reserved for the term "knowledge."

Neither is it as easy as all that for the moralist, who is likewise concerned with a constantly changing world, and who must offer advice in situations where legitimate needs of persons are in conflict with one another—as are many of the matters we are concerned with at this workshop—and in situations where the received moral wisdom of our Christian tradition does not address the topic at hand or is unclear.

I say all this as something of a warning. Many, many people have studied the human embryo, and have produced much high quality documentation of their work. Dr. Opitz's article, "Human Development: The Long and Short of It," in this volume, is an example of the collaborative efforts of highly trained, experienced observers. And many others have analyzed the findings of scientists via principles deriving from differing, sometimes contradicting, philosophical traditions.[1] And judgments about the moral "fallout" of these scientific and philosophical conclusions are made by moralists who also come from differing ethical traditions.[2] Approximate certitude may be the best we can expect.[3] *Marana tha*!

---

[1]For instance, although much Catholic material draws its inspiration from the philosophical work of St. Thomas Aquinas, the Catholic bioethicist Thomas Shannon employs Duns Scotus's doctrine of *haecceitas*, and is perforce comfortable with Scotus's doctrine of a plurality of substantial forms. On the other hand, William A. Wallace, a committed Thomist, continues to hold strongly to Aquinas's teaching that humanity in the fetus is not attained until long after conception. See his "Aquinas's Legacy on Individuation, Cogitation, and Hominization," in *Thomas Aquinas and His Legacy*, ed. David M. Gallagher, vol. 28, *Studies in Philosophy and the History of Philosophy* (Washington, D.C.: Catholic University of America Press, 1994), 173–193, also his "St. Thomas on the Beginning and Ending of Human Life," in *Sanctus Thomas de Aquino doctor hodiernae humanitatis: miscellanea offerta dalla Societa internazionale Tommaso D'Aquino al suo direttore prof. Abelardo Lobato per il suo 70. genetliaco*, ed. D. Ols (Vatican City: Libreria Editrice Vaticana, 1995), 394–407; and his "*Fides et Ratio*: The Compatibility of Science and Religion" in this volume, 155–179.

[2]A recent example here might be Jean Porter, who, though coming from the Catholic tradition, holds in her book *Moral Action and Christian Ethics*, vol. 5, *New Studies in Christian Ethics* (Cambridge: Cambridge University Press, 1995), that considering a fetus to be a human being does not entirely make it immune to abortion in some circumstances.

[3]For instance, Carol Tauer, "The Tradition of Probabilism and the Moral Status of the Early Embryo," *Theological Studies* 45.1 (March 1984): 3–33, will hold that the lack of certitude allows for certain courses of action. For Richard McCormick, the preembryo's not being a human person is "solidly probable," see his "The Embryo Debate 3: The First 14 Days," *Tablet* 224.7808 (March 10, 1990): 301–302.

The Church's Assertions

But for all that, our Church's magisterium urges a moral certitude that abortion should not be procured nor embryos experimented upon, a moral certitude that can be traced all the way back to the *Didache*, that first-century collection of Christian moral teaching, in turn claiming to depend upon the teaching and practice of the Apostles.[4] While some recent documents, such as the Congregation for the Doctrine of the Faith's 1974 "Declaration on Procured Abortion," note that there has been no unanimous tradition in the Church insisting that full human life begins at conception,[5] the same CDF's *Donum vitae* of 1987 claimed:

> The magisterium has not expressly committed itself to an affirmation of a philosophical nature [to the time of ensoulment], but it constantly affirms the moral condemnation of any kind of procured abortion. This teaching has not been changed and is unchangeable. The human being is to be respected and treated as a person from the moment of conception, and therefore from that same moment his rights as a person must be recognized, among which in the first place is the inviolable right of every innocent human being to life.[6]

Even *Evangelium vitae*, which is no less insistent than this passage, makes the demand for our treating the human embryo as a human person, but goes no further.[7] In short, the teaching of

---

[4]After the first commandment of love of God and neighbor, the *Didache* in the "Two Ways of Life and Death" used to instruct persons for baptism says, "But the second commandment of the teaching is this 'Thou shall do no murder; thou shalt not commit adultery ... thou shalt not procure abortion nor commit infanticide.'" *The Apostolic Fathers*, trans. Kirsopp Lake, vol. ii, *Loeb Classical Library* (Cambridge, MA: Harvard University Press, 1985), 311f. For the history of the Catholic Church on abortion, see John Connery, S.J., *Abortion: The Development of the Roman Catholic Perspective* (Chicago: University of Chicago Press, 1977).

[5]CDF, "Declaratio de abortu procurato," *AAS* 66 (1974), 730–747, explicitly avoids the issue of the moment of the newly conceived zygote's animation with a rational soul, because there was no constant tradition on the subject and authors disagree (738, note 19).

[6]See the CDF's "Instructio de observantia erga vitam humanam nascentem deque procreatio nis dignitate tuenda (*Donum vitae*)" *AAS* 80 (1988), 70–102, I n. 1.

[7]*Evangelium vitae* echoes the CDF document its discussion of the "unspeakable crime of abortion," (nn. 58–63), saying: "Further-

the Church's officials has been, on this matter at least, to stick with the safer course—and wisely so, given that the damage done to the embryo through experiment or abortion is irreparable and final.

But what if it can be shown that the basis for the Church's grave caution does not exist? In other words, while the Church insists that we should treat the human embryo as though it is a human person because we do not know for sure at what time, or in what way, a prehuman embryo would become a human embryo, what if our philosophical analysis of the scientific data led us to the certain conclusion that the embryo before implantation *cannot be* a human being? What then? Respected philosophers, and a goodly number of influential Catholic theologians,[8]

---

more, what is at stake is so important that, from the standpoint of moral obligation, the mere probability that a human person is involved would suffice to justify an absolutely clear prohibition of any intervention aimed at killing a human embryo. Precisely for this reason, over and above all scientific debates and those philosophical affirmations to which the magisterium has not expressly committed itself, the Church has always taught and continues to teach that the result of human procreation, from the first moment of its existence, must be guaranteed that unconditional respect which is morally due to the human being in his or her totality and unity as body and spirit: '*The human being is to be respected and treated as a person from the moment of conception;* and therefore from that same moment his rights as a person must be recognized, among which in the first place is the inviolable right of every innocent human being to life'" (n. 60, quoting *Donum vitae* I n.1). Despite its age, Guy de Broglie's paper, "Avortement et 'meurtre,'" *Doctor Communis* 27 (1974): 3–40, remains an excellent meditation upon the moral evil of abortion, even if we are not able with certainty to label it murder.

[8]A seminal text for this line of reasoning was Joseph Donceel's "Animation and Hominization," *Theological Studies* 31 (1970): 76–105. For authors influenced by this view see: Lisa Sowle Cahill, "The Embryo and the Fetus: New Moral Contexts," *Theological Studies* 54.1 (March 1993): 4–42; Clifford Grobstein, *Science and the Unborn: Choosing Human Futures* (New York: Basic Books, 1988); Norman M. Ford, *When Did I Begin? Conception of the Human Individual in History, Philosophy and Science* (New York: Cambridge University, 1988); Richard McCormick, "Who or What Is the Preembryo?" *Kennedy Institute of Ethics Journal* 1.1 (March 1991): 1–15; Thomas A. Shannon and Allan B. Wolter, "Reflections on the Moral Status of the Pre-Embryo," *Theological Studies* 51.3 (December 1990): 603–626; Carlos A. Bedate and Robert C. Cefalo, "The Zygote: To Be or Not To Be a Person," *Journal of Medicine and Philosophy* 14 (1989): 641–645. Added to this would be Jean Porter, *Moral Action,* and William Wallace (see note 1, above).

argue that we do have certitude that before implantation the embryo fails to meet the criteria necessary to being counted a human being, and they are perforce more open to certain embryological experiments and less troubled by abortifacient contraceptives. These thinkers, it is important to note, are examining the latest embryological data, using traditional philosophical and theological thoughtforms, and employing in many instances the teaching of the Common Doctor of the Church, St. Thomas Aquinas. So we are dealing here with our brothers and sisters in the faith, who wish, like us, to see the truth be the basis for our action or inaction. While one surefire way to address the claims of these writers would be to demonstrate beyond a shadow of a doubt, with scientific certitude, that the newly conceived zygote can only be a full member of the human family, such a demonstration is likely not in the offing, as experience and official church documents indicate. However, if it can be shown that the arguments they employ to question the caution of church officials have scientific and philosophical difficulties, possibly insuperable difficulties, then at the end of the day we are left with the Church's wise admonition that love of neighbor includes love of the neighbor who might not yet be there. So my goal in what follows is admittedly negative: to show that the bases upon which these thinkers raise doubts about the elements of the Church's caution are flawed and profoundly so. The result will be, if anything, that the Church's admonition is not only not illfounded, but is the best practical conclusion available in light of what we do know, and do not know, and is the most generous and liberal call for the respect of others' basic rights to life.

Our Tradition on Personhood

The Catholic position on personhood is first and foremost an "entitative" position; that is, we Catholics consider someone to be a person who is, simply, a living member of the human species, regardless of whether that individual can, right here and now, perform functions that are exclusive to the human species, such as intellective reasoning. This contrasts with what might be called a "functional" account of personhood, according to which an individual is to be counted a person when he can do something we consider the hallmark of being human, such as reasoning, or all too often, being "self-aware"—a holdover from the view that knowledge of self is prior to knowledge of other things, which in turn would postpone the granting of personhood in children to about seven years of age.

So when we Catholics examine the activity of the human embryo relative to its being a person, we are really asking the question "is the embryo a living member of the human species?"—a question that quietly employs the classic theological definition of person, coming from Boethius: is the embryo an "individual substance of a rational nature"? The answer, from theologians ranging from Norman Ford to Richard McCormick to St. Thomas, is "no." Those reading and employing contemporary biology will say that the embryo fails to meet the test of being an "individual substance," and those depending upon St. Thomas will say that, even should we grant that the embryo is some kind of a substance, it is not a human kind of substance, at least not yet. At the risk of serious oversimplification, I shall base my subsequent remarks upon these two areas of concern: the substantial unity of the embryo, on the one hand, and its being a human life, on the other.

The Embryo's Substantial Unity

A few years back I wrote some articles on these issues in the journal *Theological Studies*,[9] believing then as now that contemporary authors who claim that the embryo is not an "individual substance" mistakenly elevate a minor biological fact into the main principle, the *crux interpretatum*, of all their consideration of biological data. The biological fact in question is what is called "totipotency." What is the problem? Traditional Catholic accounts of the entitative constitution of human persons share two allied assumptions: first, that one's personhood is incommunicable to others; and second, that God cannot infuse a single rational soul into two or more distinct bodies. Embryology indicates, however, that the newly conceived zygote can be the source of more than one human being—as occurs in identical twins, and even in quadruplets—and that, further, until about two weeks after fertilization, each individual cell of the preimplantation embryo has the root capacity (totipo-

[9]See my "Delayed Hominization: Reflections on Some Recent Catholic Claims for Delayed Hominization," *Theological Studies* 56 (1995): 743–763, which was immediately followed in the same issue by Jean Porter's "Individuality, Personal Identity, and the Moral Status of the Preembryo: A Response to Mark Johnson," *Theological Studies* 56 (1995): 763–770. Thomas Shannon later responded in his "Delayed Hominization: A Response to Mark Johnson," *Theological Studies* 57 (1996): 731–734, and I finally responded to him with "Delayed Hominization: A Rejoinder to Thomas Shannon," *Theological Studies* 58 (1997): 708–714. Much of what follows derives from that last article.

tency) to become a whole human being. It seems therefore that from fertilization to implantation the embryo has a capacity to "share itself around," thereby failing this test of incommunicability. And the totipotency of the cells correlatively suggests—because they can each produce a whole—that they are more wholes than they are parts-of-a-whole. It just does not seem that before the fourteenth day we have a single entity, an individual.

This is the view first of Norman Ford, whose very influential book, *When Did I Begin?* [10] has created a lexicon for considering this topic, a lexicon employed by Thomas Shannon[11] and others. For Ford and those who follow him there is a distinction between the embryo's "genetic uniqueness" and its "developmental individuality." Genetic uniqueness results from one's unique human genome, present from conception, which, while sufficient to classify the embryo as pertaining to the human species, is not sufficient to guarantee that it will be one, rather than many, human beings; identical twins are genetically the same, but are not the same individual. So what we must await is that state of the embryo at which its many cells have become incapable of producing another distinct organism. This occurs when the cells have undergone "restriction," such that, e.g., this particular cell can only be a liver cell. At that point the concrete has set, as it were, and the embryo is considered a developmental or ontological individual.

Now, I believe that insufficient attention is being given to the multitude of biological activities the preimplantation embryo performs. My study of the data (including the work of Dr. Opitz and Scott Gilbert) leads me to conclude that there is a genuine biological, living unity in the embryo from conception

---

[10]Norman Ford, *When Did I Begin?* The book received some criticism; see Nicholas Tonti-Filippini, "A Critical Note," *Linacre Quarterly* 56.3 (August 1989): 36–50, to which Ford replied in "When Did I Begin—A Reply to Nicholas Tonti-Filippini," *Linacre Quarterly* 57.4 (1990): 58–66; Anthony Fisher, " 'When Did I Begin?' Revisited," *Linacre Quarterly* 58.3 (1991): 59–68; id., "Individuogenesis and a Recent Book by Fr. Norman Ford," *Anthropotes* 7.2 (1991): 199–244; Paul Flaman, " 'When Did I Begin?' Another Critical Response to Norman Ford," *Linacre Quarterly* 58.4 (1991): 39–55; and Ronald K. Tacelli, "Were You a Zygote?" *Josephinum Journal of Theology* 4.1 (Winter/Spring 1997): 25–36.

[11]See also Shannon's "Cloning, Uniqueness, and Individuality," *Louvain Studies* 19 (1994): 283–306; and "Issues and Values in Genetic Engineering: A Survey," *Chicago Studies* 33 (1994): 196–204, both of which contain the same doctrines he holds here.

forward, and that those who employ this genetic/developmental distinction do not really think that the preimplantation embryo is a living thing, an organism. The life of a biological reality is its ontology, so I think that one is compelled to hold either that the embryo is a living organism, or that it was no thing at all, but a cluster of things unified only by contact, what Aristotle calls "a heap."[12] Despite the earnest efforts of writers such as Thomas Shannon, it seems to me that writers who follow the teaching of Norman Ford will fail to recognize the full biological unity of the embryo, because they instead fix upon the so-called "totipotency" of the preimplantation embryo's cells; this in turn renders flawed their account of what a biological part in this organism might be, an account that prevents them, in my judgment, from seeing its full ontological status.

For followers of Ford, the port of entry into this discussion—and indeed the reason to affirm a distinction between genetic and developmental/ontological individuality—is the so-called totipotency of early embryonic cells, a biological fact that serves as their single principle of interpretation for all other data concerning the preimplantation embryo. We all agree, because we must agree, that totipotency is a fact; but our interpretations differ. And in my experience whenever I see emphasis upon totipotency, I am sure that subsequent discussion will minimize the biology operative in that multicelled organism that is the embryo.[13]

In the early embryo the cells of which it is composed replicate the DNA contained in their nucleus, and, when that is complete, cleave themselves so that a half-size copy of the original is made, nucleus and all, and the original is now half of its earlier size. The DNA resident in the cell's nucleus remains fairly constant through these early cleavages, so that the DNA in a cell at, for example, the sixteen-cell stage, mirrors that of the DNA that was in the nucleus of the single-celled zygote—if I should be proven wrong here that only helps my argument. Over time, certain genes in the DNA are switched on and off in

---

[12]See Aristotle, *Metaphysics* 7.17 (1041b11–33) and Aquinas's commentary *In VII Metaphysicorum*, lect. 17, nn. 1672–1680 in *XII Libros Metaphysicorum*, ed. M.-R. Cathala and R.M. Spiazzi (Turino: Marietti, 1977), 398–399, where he explains what a *cumulus* (a "heap") is.

[13]Shannon proposes to present the process of embryogenesis, but speaks almost exclusively of totipotency and not about intercellular communication, mutual regulation, etc. in "Cloning, Uniqueness, and Individuality," 285.

the process of restriction, the goal of which is to produce tissues of cells fitted to certain functions (e.g., liver, heart, arterial tissues). Those cells in turn produce other cells containing their restrictions (certain genes switched on and off), but will sometimes add new restrictions that will be passed down to their descendants, in cascading restriction. This process is gradual, however, because of the embryo's important need for regulative development. In this type of development, predominating in vertebrates, an individual cell's fate is determined primarily by its interaction with other cells around it, which influence it to modify its DNA in certain, specified ways. Regulative development differs from the mosaic development of invertebrates, whose cells' characteristics are defined more by the nonnuclear factors of a cell (e.g., the cytoplasm, which is successively partitioned during cell cleavage). In such a case the loss of a particular cell can mean that the cells that would descend from it, and their corresponding functions, would be entirely absent in the organism (e.g., missing structures, parts of tail or wings, etc.). But in regulative development the loss of a particular cell simply means that the remaining cells that were with the now-lost cell will alter their own cell fates—which they can do because their DNA has not yet been irreversibly restricted—and can themselves provide for the DNA specification and functions that the missing cell had once provided. These cells (blastomeres) are able to take full advantage of the genetic information they possess, and, working in concert with the other cells that constitute the embryo, insure its self-regulation. This is the primary sense of totipotent; these cells have the root ability to fulfill any other cellular functions in the embryo of which they are a part.

A side effect of this ability to aid in regulation is that, if an early cell should be removed from the embryo and be separated from the influence of other cells, it is able to make full use of its genetic information and to produce a completely other embryo! Because of this characteristic such cells are said to have "totipotency" in a second sense; they are able, on condition of separation, to produce another whole.

It is crucial to emphasize that this latter totipotency of a cell in an early embryo to become an entirely other embryo is *conditional*. To speak of it with no qualification implies that such a cell could activate its own potency entirely from within, and at any moment—which is not the case here. So I think it best to term the early embryonic cell, somewhat paradoxically, as "potentially totipotent," because that indicates that its ability to become a full-blown embryo is dependent upon an external change that results in a new state of affairs for it. In short, the cell must be physically removed from the embryo in a laboratory, or through some accidental means (e.g., via a spontaneous genetic mutation preventing intercellular communication or being dislodged during the hatching from the zona pellucida, as some speculate is a cause of twinning).

It is true to say that "these cells can become other organisms," or to say "that such a capacity [to become another organism] actually resides within the organism."[14] But I would argue that it is crucial to delineate the character of those capacities, determining what comes from within and what comes from without, for these are signs of the thing's status as a self-standing entity. "To-be-able" has many possible meanings. Clay is able to become a pot (a passive potency). A two-year-old human child is able to play the piano—both a passive potency (he must be taught) and an active potency (he will instill the habit through practice). Yo-Yo Ma is able play the cello (a habitual active potency). And a rock I hold over my head can fall when I let go (an active potency awaiting only the removal of a prohibition by some outside force).

But the early embryonic cell seems to be a different case. In one way it is like the rock just mentioned in that its ability to grow into a whole other embryo depends upon the removal of some outside, prohibiting influence. In another way, however, it differs because while it is a functioning part of the embryo it is not working to actualize that ability to become another embryo; this is in contrast to the rock, whose heaviness indicates that it "works" to get down to the ground, even when held up. In short, the early embryo's cell is not "chomping at the bit" to be free of the other cells so that it can realize its ability to become a whole other embryo. While part of the embryo, it receives messages from other cells and reacts to them in an appropriate way and sends messages to them to regulate their activity. The early embryo's cells are other-oriented, heterogeneous, contributing parts of the whole to which they belong.

I say this for two reasons. First, it is characteristic of individual entities, of substances, to initialize the realization of their potencies or abilities without the aid of some external thing. Thus the early embryo's cells should not really be con-

---

[14]Shannon uses the same language in speaking of the early embryo and the eventual time of its restriction: "Any of its individual cells until that time can be a whole other being." Ibid., 301.

sidered "individuals" in any full sense of the term when they are contributing parts of the embryo. Second, we need to guard against inadvertently taking this "potential totipotency" of the cells and attributing it to the whole early embryo. If we do, then we are regarding the early embryo as a homogeneous entity; the character of the parts can be fully predicated of the whole, and vice versa. The parts are apt to become other organisms, and the whole is apt to become other organisms. And so we lose sight of the intrinsic dynamism of the parts relative to one another, and focus instead upon the prospect of the embryo's becoming something other than what it is.

Focusing on the early embryo's cells in this uniform way sets the stage for Ford's and Shannon's thinking about its ontological status as a whole. Failing to note the cell's character as an ordered, contributing, heterogeneous part of the early embryo, a cell whose totipotency is checked and conditioned by the others that surround it (cells whose own totipotency that cell itself checks through its own influence upon them) then it is no wonder that they fear for the whole embryo's durability or staying power; for, since they take "totipotency" to mean "at-any-moment-capable-of-producing-another-organism," their biological account has the preimplantation embryo in immanent danger of spontaneous disintegration.[15]

This section can be summarized under two points. First, an organism's life and its need to interact with its environment are the reasons for all its developmental structures and characteristics, and biological data need to be read in that light. Second, it is a life-serving, biological advantage for the early embryo to be comprised of cells that perform specific functions, but perhaps other functions as well, should the need arise, occasioned by the damage that the loss or death of other cells inflicts upon the whole. The totipotency of the early embryo's cells works, therefore, to preserve the living unity of the whole embryo, and as such, cannot possibly be a reason for denying that unity from the outset. So what we are left with is an organized body of heterogeneous parts influencing, and influenced by, each other, which works to preserve its existence and enhance its ability to interact with its environment—in that sense, an organism like us.

---

[15]Shannon claims that the early embryo is "biologically indifferent to singleness until after restriction occurs," ibid. Would it be as normal for it to become many as it would be for it to become one?

St. Thomas and the Embryo

It could be the cause of some mirth to note that, of all of St. Thomas's doctrines, it is his doctrine of the "delayed hominization" of the human embryo that philosophers and theologians are clamoring to defend or appropriate to the present setting. And the reason for this oddity is, of course, that Thomas was writing in a time when the science was crude, and little was known of the workings of the human body that we now understand by virtue of the microscope. And what little Thomas knew about embryology he got from the doctrine of Aristotle, whose doctrine was already fifteen hundred years old—hardly a reason for confidence. Yet the fact is that a presentation of the doctrine of Thomas and Aristotle sounds, in the main, strikingly similar to what we know to be sequentially true about human embryological development, which is precisely why modern authors, starting with the outlines of contemporary embryology, and then reading Thomas's own, and the philosophical principles he uses to elucidate and defend his position, see Thomas's thought as a useful tool in describing our current knowledge. In short, we take Thomas's philosophy, tidy up the details of his embryology, and then we are ready to address the issue of hominization with many arrows in our quiver.[16]

So what was the teaching of Aristotle, which Thomas used in his work?[17] Aristotle was a staunch opponent of the theory of "preformation," according to which the male's semen contained a miniature human being, ready-formed, and which just needed an environment in which to grow. But having abandoned that embryological option, he now had the more difficult task of explaining how humans came to be, on the basis of the physical evidence he had available to him, which offered little in the

---

[16]For instance, Jean Porter, in *Moral Action and Christian Ethics*, claims that "as far as his biological presuppositions go, Aquinas was essentially right," 122. My claim below will be the exact opposite. See also Robert Pasnau, *Thomas Aquinas on Human Nature: A Philosophical Study of Summa theologiae 1a, 75–89* (Cambridge: Cambridge University Press, 2002), 100–130.

[17]The best exposition of Aquinas's thought on this issue remains that of Michael Allyn Taylor, in his 1982 dissertation *Human Generation in the Thought of Thomas Aquinas: A Case Study on the Role of Biological Fact in Theological Science* (Ann Arbor: UMI Dissertation Service, 1997). In what follows, however, I am basing my claims on the reading of Thomas's texts themselves especially *Summa theologiae*, I, Q. 118.1 and 2.

way of detail. But Aristotle was pretty sure of this: the male's contribution, the semen, was small in terms of quantity, and was of course dwarfed by the size of the child that is born at the end of gestation; so he figured that the semen's function in reproduction was not to provide the material out of which the fetus was constructed. Rather, the male's semen was active, while the female's role was passive, for she has within her the menstrual blood which, if it is not fertilized by the male seed, is spent every month. Both of these elements of reproduction, it is important to note, are for Aristotle the result of partial digestion of food; in other words, the semen and the menstrual blood are not, strictly speaking, "parts" of the body, or even remnants of parts of the body—this will be important to us later. The female's menstrual blood is worked over by the male seed, somewhat like, Aristotle notes, the way in which rennet curdles milk.[18]

This last image brings us to two other features of Aristotle's teaching. First, Aristotle held that the menstrual blood that the semen works and activates exists in a very, very low grade of organization—it was, he thought, a type of digested food—such that much time and effort will be required to bring this material from its early blood-state to human material; and so we are inevitably speaking about a passage of some time between fertilization and the development of a living thing, much less a human living thing. Second, because Aristotle had figured that the female's contribution was passive in the process of reproduction, and not an active participant in the development of structures, tissues, and organs, it must be the case that the male semen remains separate from the female contribution during the process of development, constantly working on it as a separate agent. So the process of human development proceeds from a minimally organized state to a progressively more organized state. But even here the emerging fetus is not the principal agent of change in its development, for no passive thing produces in itself the changes necessary for it to migrate from one level of substantial life to another. That is accomplished by the male seed, acting as a kind of separated instru-

---

[18]The principal source for Aristotle's teaching on embryological development is his work *On the Generation of Animals* (*De generatione animalium*), especially book 1, ch. 21 and following (720a25–). For more detailed presentations of Aristotle's teaching and its use in medieval Christian and Muslim thought, see *The Human Embryo: Aristotle and the Arabic and European Traditions*, ed. G.R. Dunstan (Exeter, Devon: University of Exeter Press, 1990).

ment of the soul of the human father, intending through its action to produce another human being like the father—which is why, incidentally, the cause of gender differences for Aristotle arises not from the side of the male semen (which intends a male, since "like produces like"), but rather from the side of the maternal material contribution, or other external factors.

So Thomas, adopting the embryological doctrine of Aristotle in a Christian context, a context that holds that the spiritual human soul, being the form of the human body, must be the direct creation of God, followed the Greek philosopher in holding that the process of human reproduction took place over a period of some time, because of the lack of organization of the matter of the embryo, and because they thought that the semen's continual separation from the fetus meant that it was the cause of the developmental changes that take place in the embryo, and not that the embryo was acting upon itself to produce those changes within itself. Only when the embryo had been brought to animal life, capable of producing in itself the organs of sensation, or actually having them, could the embryo receive a specially created human soul.

Let me make a comment about St. Thomas's conception of the human soul. Although he held, following Aristotle, that the development of the human embryo brought it first from a stage of no life at all to a nutritive stage (nutrition and growth), and then second to an animal stage (sensation and movement), and then finally to the third, fully human stage of a rational life, he did not therefore think that the human person is endowed with three continuous principles of lifeactivities: a nutritive soul, together with an animal soul, together with a rational soul. This would amount to holding a doctrine of the "plurality of substantial forms," a doctrine that Thomas vigorously opposed from the beginning of his writing career to its end. Indeed, his opposition to this doctrine cost him dearly after his death, as his writings were condemned. What Thomas held instead was the theory that the animal soul performed all the functions that are characteristic of an animal (sensation, memory, local motion) and performed as well the functions that are performed by the soul of a plant (nutrition and growth), while remaining the one soul of the whole animal. The same holds true of the human rational soul—which, unlike the animal and vegetative souls, is fully immaterial and immortal—for it performs the functions associated with human intellection (principal among which are direct knowledge of material things in an immaterial way and human willing), and performs as well the functions that the other "lower souls" provided: nutrition, growth, sensation, local motion. So when Thomas says that the human soul contains *virtually* the lower souls, he does not mean that it is possible for the human rational soul to be "peeled away," as it were, leaving behind an animal soul, or that the remaining animal soul in turn could be "peeled away" to leave behind the vegetative soul. Rather, the one human soul, being rational and animal and vegetal, performs all these functions for the one human body.[19] How this effects his discussion of human embryology in the present situation I shall now suggest.

St. Thomas and What We Know Now

When scholars speak of Thomas's use of embryology they regularly say he has the biology wrong, but that he has the philosophy right. We should not therefore try to salvage his biology, the claim goes, but we can apply his philosophical reasoning about errant biology to current, demonstrated embryology. As a committed student of Thomas's writings I must confess that that none of his efforts on this issue is salvageable. It is simply not possible to "tweak" what Thomas has said, for the very simple reason that the entire embryological basis upon which Thomas's argument rests, and from which the trajectory of his later claims about delayed hominization takes its origin, is factually false, and completely so. Thomas thought that the male's seed did not become one with the female's material contribution, which he believed was in such a crude state that effort was needed to bring the maternal contribution to even a minimal state of life. That is false, for we know now that the sperm and the egg prior to fertilization are highly and equally organized, though "half-organized," as it were, such that the union of the two at fertilization results in a single body that is materially one, and formally or organizationally one, and is indeed organizationally more sophisticated than either the sperm or the egg had been, on it own, prior to fertilization. After fertilization occurs, of course, it is improper in the discipline of biology to speak further of "the sperm" or "the egg," since those two erstwhile realities no longer exist, but have become "parts" of another whole. Thomas thought that the embryo was the passive recipient of the semen's agency, since the semen worked-over, and shaped, and "built" the structures in the passive embryo, first developing in it the structures necessary for nutrition, then eventually sensation, and so on. That is false, for we

---

[19]See *Summa theologiae*, I, Q. 75.3 and 8.

know now that the embryo has within itself the formal capacity in its DNA to implement the structural "design" of a human being from its outset, needing nothing outside of itself to do this save nutrition. We know that it is its own agent of change and development, with one part of the embryo working on another part, and so on, as it constructs within itself both the organs and systems necessary to life outside of the womb, and, importantly for our consideration of specifically human development, the organ the human intellect and will use in their activities, namely the human brain, and particularly the upper brain.

So how would we describe the newly-conceived zygote or embryo to Thomas? We would say that at human fertilization a new, single-celled entity comes into existence—the sperm and egg cease to be—and that the organizing source in this entity (the nucleus and its DNA) cannot enjoy a state of higher organization and sophistication than it currently enjoys. Indeed, all subsequent development in the embryo's body is accomplished not by adding sophistication to the DNA, but rather by diminishing the potency of cellular DNA through gene-switching. Hence, there is no need to insist upon a lag of time in order to arrive at an intrinsic principle of order in the embryo. We would also say that this entity is the agent of the growth and developments of its parts, using its internal formal plans and its internal organs and structures to cause the formation of these parts. This is something we know to take place in established humans (e.g., cerebral and sexual development after birth, which of course we attribute to the human body, and not to an outside source).

How would Thomas respond? This is of course a hypothesis, but, because he does hold to the doctrine that the one human, rational soul performs all the functions of all the various structures or "layers" of human existence, including those involved with growth, I think that he would have to admit that there is no reason why we should not hold that the one human, rational soul is once and for all the continuous life principle of this highly organized, self-building organism that we know to come into existence at fertilization. To suggest that the human embryo is first alive with a subhuman animal soul, to be replaced by a human soul at a later time, when that solitary human soul could continuously cause all the functions in the embryo from the outset, would be, I think Thomas would say, to posit an unnecessary step in God's necessary interaction with the natural world. God could, of course, perform that "stutter-step," but from the basis of the physical evidence available to observers of the embryo, we would not know when that step took place, and would therefore be ill-advised to act on the assumption that such a "step" is taking place at all.

The Humanity of the Embryo

I have endeavored to show here that the central arguments backing the claim that we can be certain that the early human embryo is *not* a person in the way recognized by our Church's intellectual tradition are radically flawed. First, I have worked to show that arguments against the unity of the embryo are made only when one fails to let the biology of the embryo speak for itself, and one instead imports a doctrine of totipotency so imperialistic that all other facts about the biological unity of the embryo fade from view. Second, I have argued that recourse to the doctrine, and even arguments, of St. Thomas on this issue are gravely problematic, since we know at least with certainty that the biological rug has been pulled out from under Thomas's feet. His authority and argumentative strategy on this issue are best left behind.

I began my comments with the caution that "approximate certitude may be the best we can expect" in this matter; and I do believe that to be the case. There is no reason that the human embryo cannot be a full, living member of our human family. It is just that the disciplines of biology and contemporary philosophy do not address the issue in a way that will satisfy all ethicists, especially with pluralism reigning supreme on all sides. I do believe that the arguments for the humanity of the embryo are sufficiently strong, and arguments against it so increasingly weak, that within our Catholic community there is every reason to support vigorously the contention that the embryo's life should be thought of and treated as a life like our own. But when it comes to treating others humanely, we all know that is it our conduct as much as our doctrine that spreads the Gospel of Life. *Marana tha*, indeed.

Disputed Questions

THE MORALITY OF "RESCUING" FROZEN EMBRYOS

WILLIAM E. MAY

Thousands of tiny unborn children, produced through in vitro fertilization, have been frozen and put into what the late, great French geneticist Jerome Lejeune aptly termed "concentration cans." Many of these frozen embryos have been abandoned by those who ordered their "production," and the question thus arises, "Can anything be done to save them?" A lively debate is currently being carried out by Catholic moral philosophers and theologians—and others—over the morality of "rescuing" these tiny orphans by having them transferred from their concentration cans to the wombs of women who volunteer to protect and save their lives.

I will defend the position that it is *not* intrinsically evil for a woman, whether married or single, to choose to have a frozen human embryo who has been orphaned by those who brought him or her into existence transferred from the freezer into her womb as a means of protecting and saving this unborn child's life.

I will proceed as follows. First, I will briefly summarize the teaching of the Church on the morality of human acts and show that some kinds of human acts, specified by their object of choice, are *intrinsically evil* and can never be morally justified, and that among such acts are the laboratory reproduction of human persons, serving as a surrogate mother, and sexual union outside of marriage. Second, I will develop an argument to show that the *human act* freely chosen by a woman to have a frozen embryo transferred from its concentration can to her womb in order to protect and save his or her life is in no way intrinsically evil. Third, I will propose that, although it is not intrinsically evil for an unmarried woman to save the life of a frozen embryo in this way, it is preferable that a married woman do so, and that she and her husband serve as this child's adoptive parents. Fourth, I will consider certain *caveats* that need to be seriously taken into account in order to prevent or at least lessen the possibility of gravely reprehensible consequences that might result from legitimate efforts to save the lives of frozen embryos. Finally, I will respond to the major arguments advanced by reputable philosophers and theologians to support their claim that "rescuing" frozen embryos in this way is intrinsically evil, and therefore, cannot be morally justified.

The Morality of Human Acts

Here I will follow the teaching on the morality of human acts set forth by Pope John Paul II in his encyclical *Veritatis splendor,* a teaching rooted in the Catholic tradition. According to John Paul II, "the morality of a human act depends primarily and fundamentally on the 'object' rationally chosen by the deliberate will" (n. 78). In a very important passage, well summarizing the Catholic tradition as expressed by St. Thomas, whose authority he invokes,[1] John Paul II then says:

> In order to be able to grasp the object of an act which specifies that act morally, it is therefore necessary to place oneself *in the perspective of the acting person.* The object of the act of willing is in fact a freely chosen kind of behavior. To the extent that it is in conformity with the order of reason, it is the cause of the goodness of the will; it perfects us morally.... By the object of a given moral act, then, one cannot mean a process or an event of the merely physical order, to be assessed on the basis of its ability to bring about a given state of affairs in the outside world. Rather, that object is the proximate end of a deliberate decision

---

[1]He refers to St. Thomas Aquinas, *Summa theologiae,* I-II, Q. 18.6.

[free choice] which determines the act of willing on the part of the acting person (n. 78, original emphasis).

John Paul II's thought, as noted already, is rooted in St. Thomas's analysis of the morality of human acts. Thomas sharply distinguished between the "natural species" of a human act, defining it as a "process or an event of the merely physical order," for instance, sexual coition, and the "moral species" of a human act.[2] Human acts, precisely as "human" and "moral," do not have "forms" given to them by nature; rather these forms, putting them into their "species" as "good" or "bad," are given to them by reason, which can discern the "objects" or "ends" specifying them.[3] The "object" of the human act is precisely *what* the agent freely chooses to do. It is precisely what can be called the "intentional content" of one's free choice[4] or what Thomas called the "subject matter" or *materia circa quam.* Thus the "object" of the human act of adultery is precisely "to have sexual coition with someone other than one's spouse." This object is precisely what John Paul II, in the passage cited above, called the "proximate end of a deliberate decision," it is "the freely chosen kind of behavior" *willed* by the acting person. St. Thomas put the same truth somewhat differently, but precisely, when he said: "moral acts receive their 'species' [either good or bad] according to what is *intended* [either intended as the *end* for whose sake one acts or chosen as the *means* toward that end], not by what is outside the scope of one's intention."[5]

In other words, human acts are specified primarily by *what* the acting person *chooses to do,* by the intelligible proposal that he adopts by choice and then executes, by the *intentional content* of his choice; and this object is the *proximate* end of the human act in question, although the act can be subordinated to *further, remote ends.* But good remote ends can in no way justify acts specified by a morally bad object of choice.

---

[2]St. Thomas Aquinas, *Summa theologiae,* I-II, Q. 1.3, reply 3.

[3]Ibid., Q. 18.10.

[4]On this see the insightful essay of Martin Rhonheimer, "Intrinsically Evil Acts and the Moral Viewpoint: Clarifying a Central Teaching of *Veritatis splendor,*" in *Veritatis Splendor and the Renewal of Moral Theology: Studies by Ten Outstanding Scholars,* eds. J. A. DiNoia, O.P., and Romanus Cessario, O.P. (Chicago: Midwest Theological Forum, 1999), 161–194.

[5]See St. Thomas Aquinas, *Summa theologiae,* II-II, Q. 64.7: "actus autem morales recipiunt speciem secundum id quod intenditur, non autem ab eo quod est praeter intentionem."

Given this understanding of "object," it is then easy to grasp the Holy Father's argument regarding the morality of human acts, which he summarizes by saying, "reason attests that there are objects of the human act which are by their nature 'incapable of being ordered' to God because they radically contradict the good of the person made in his image" (n. 80).

What this means is that a human act is morally bad *if* the object freely chosen by the acting person is precisely to damage, destroy, or impede a good of the human person—a truth emphasized by John Paul II earlier in *Veritatis splendor* when he affirmed that the precepts of the Decalogue concerned with our neighbor—"you shall not kill," "you shall not commit adultery"—although negatively expressed, are meant to protect the inviolable dignity of the human person made in God's image, and that they protect this dignity by protecting his *good,* at the level of the many different *goods* that characterize his identity, goods such as human life itself, the communion of marriage, etc. (cf. nn. 12–13).[6]

Human acts specified by objects that cannot be referred to God because they violate the good of the person made in his image are intrinsically evil. Among acts so specified are the intentional killing of innocent human persons, adultery or the destruction of the communion of marriage, making babies in the laboratory—an act that treats them as *products* inferior to their producers and not, as they really are, *persons* equal in personal dignity to those who beget them—and acting as a "surrogate mother" or as one who gestates a child in her womb precisely for the benefit of other persons who wish to "have a child." Intrinsically evil too are human acts in which human persons freely choose to generate human life outside of the marital act or to violate their marital communion by intentionally getting pregnant by conceiving a child outside of marriage. I will not attempt here to show the truth of all this, but will take it as a given. From this it logically follows that a woman cannot "rescue" a frozen embryo if the precise object of her choice is to generate human life outside the marital act, to serve as a surrogate mother, or to get pregnant by conceiving a child outside of marriage.

---

[6]On this see St. Thomas, *Summa contra gentiles*, Book 3, ch. 122: "God is offended by us only insofar as we act contrary to our own good" (*Non enim Deus a nobis offenditur nisi ex eo quod contra nostrum bonum agimus*).

Transferring Frozen Embryos Is Not Intrinsically Evil

I believe that the most important arguments to show that it is not intrinsically evil but rather morally good for a woman to choose to have a frozen embryo transferred from its "concentration can" to her womb in order to protect and save its life have been developed by Germain Grisez and Helen Watt. I will first summarize Grisez's line of reasoning, then that of Watt, and offer a summary conclusion.

Grisez presents his argument in answering a question raised by a single woman whose sister had committed suicide after her husband abandoned her for his secretary, leaving behind a frozen embryo. The woman, firmly opposed to in vitro fertilization, surrogacy, and all forms of artificial reproduction, wanted to act in conformity with magisterial teaching and wondered whether she would be acting in a morally upright way if she had the frozen embryo transferred to her womb, nurtured and cared for it until birth, and then gave it up for adoption.

Grisez clearly distinguishes her *object of choice* from that of those responsible for producing the child immorally through artificial means. The object of her choice is precisely *to transfer the frozen embryo to her womb*—and I will take up this aspect of Grisez's argument more fully below. He likewise points out that the woman is "proposing to bear the child not on anyone else's behalf but simply to save his or her life." Thus, if her proposal succeeds, and even if she gives the child up for adoption after delivery, Grisez holds that she is "not ... acting as a surrogate mother" but is rather "acting in much the same way as a mother who volunteers to nurse at her own breast a foundling conceived out of wedlock, abandoned by his or her natural mother, and awaiting adoption by a suitable couple. Like that mother's nursing, the nurture [she hopes] to give will in no way involve [her] in the wrongs previously done to the baby and will be offered to him or her for his or her own sake, not done as a service to anyone else. Therefore," Grisez concludes, the woman "can certainly try to save the baby without acting contrary to what the Church has taught regarding IVF and surrogate motherhood."[7]

Grisez further points out that the *end* for whose sake the woman chooses to have the frozen embryo transferred into her womb is good, namely, to attempt to save the baby's life; and the

[7]Germain Grisez, *Difficult Moral Questions*, vol. 3, *The Way of the Lord Jesus* (Quincy, IL: Franciscan Press, 1997), 241.

means chosen to pursue this end is unambiguously *to transfer the embryo from the freezer to her womb*. He summarizes his analysis of the chosen means very clearly in the following passage:

> Nurturing the baby in your womb surely will not be wrong; if someone transferred an embryo to your womb without your consent, abortion would be wrong, and it would be your duty to nurture the baby, just as it is the duty of a woman who has been raped and finds herself pregnant. Thus, if anything makes your project intrinsically wrong, it must be having the embryo moved from the freezer into your womb. But this is not at odds with any basic human good. It protects life rather than violates it; since the new person already exists, it does not violate the transmission of life; and it has nothing to do with the good of marriage, because it is not a sexual act, and the relationship between you and the baby is neither marital nor a perverse alternative to the marital relationship.[8]

Some might object (and, as we shall see, some *do* object) that if a woman intentionally allows herself to become pregnant by means other than engaging in the conjugal act with her own spouse, she is doing something intrinsically immoral. Grisez's defense of the moral goodness of a woman's choice to have an embryo transferred to her womb as the chosen means of protecting and saving its life addresses this potential objection *indirectly* when he affirms that her act "has nothing to do with the good of marriage, since it is not a sexual act and the relationship between you [the woman] and the baby is neither marital nor a perverse alternative to the marital relationship." This objection is more directly answered, in my opinion, by Helen Watt. Hence I now turn attention to the reasoning she uses to justify rescuing frozen embryos in this way.

Watt does not make use of the Thomistic distinction between the *object chosen* and thus *intended* (which gives, as we have seen, the act its *moral species*) and the foreseen consequences of one's choice—consequences that, as St. Thomas rightly emphasized, "fall outside the scope of one's intention,"[9] and which do *not* give the act its moral species. She nonetheless, so it seems to me, makes the same point in her analysis of what the woman is doing in *allowing herself to become pregnant* by having the embryo transferred from the freezer to her womb. Watt notes that the expression " 'allowing oneself to be made pregnant' covers two quite dif-

---

[8]Ibid., 242.

[9]See St. Thomas Aquinas, *Summa theologiae*, II-II, Q. 64.7, text cited in note 5 above.

ferent intentions, which affect in different ways the morality of what is being done." Continuing, she presents a clear analysis of these "two quite different intentions":

> The first intention is to allow a child *to come to be*—to be created—inside one's body. Such an intention can exist in combination either with the intention to have intercourse or with the intention to have artificial insemination. Intercourse and AID or AIH are different ways of achieving the result that a child is created inside the body of a woman. For this reason, artificial insemination can be seen as wrongfully displacing the marital act.
>
> Quite different from the intention to have a child *come to be* inside one's body is the intention to have a child *put* inside one's body.... It is ... not the case that uterine pregnancy—that is, pregnancy after implantation—is directly caused by intercourse. If this is so, should it still be said that intercourse must *always precede* uterine pregnancy? What I want to argue is that whereas ideally intercourse *should* precede uterine pregnancy, the only *absolute* moral requirement is that [marital] intercourse should precede—and indeed directly cause—*in vivo* conception.[10]

I would express Watt's argument in more Thomistic terms by saying that it is intrinsically immoral if one freely chooses [intends] to have a child come to be in one's womb or to become pregnant as a result of nonmarital intercourse and/or artificial insemination or in vitro fertilization. The object specifying the act is intrinsically immoral. But a woman who freely chooses to have an already existing child put into her womb, that is, a child *that she has not conceived* by nonmarital intercourse or artificial insemination/in vitro fertilization, engages in an act specified by a different moral object, namely, to transfer an orphaned frozen embryo into her womb. Although she is made pregnant as a result or consequence of this choice, the object of her choice is *not* to become pregnant outside of marriage and the marital act and in this way to abuse her sexuality.

In my opinion Watt's argument complements Grisez's, expressing the same main point somewhat differently. For both Watt and Grisez the *object* morally specifying the act of a woman who seeks to rescue an abandoned and frozen embryo by having it transferred to her womb is not intrinsically evil and quite

---

[10]Helen Watt, "Are There any Circumstances in Which It Would Be Morally Admirable for a Woman to Seek to Have an Orphan Embryo Implanted in Her Womb?" in *Issues for a Catholic Bioethic,* ed. Luke Gormally (London: The Linacre Centre, 1998), 349–350.

different from the *object* morally specifying the act of a woman who chooses to become pregnant outside of marriage as a result of artificial insemination, in vitro fertilization, fornication, or adultery. A woman who allows herself to become pregnant as a result of having an already existing child orphaned by its parents transferred into her womb from the freezer does not have as her moral object *becoming pregnant outside of marriage.* Her becoming pregnant is a foreseen and inevitable result of having the embryo transferred into her womb.

Better for Married than Unmarried Women

If the woman who chooses to rescue a frozen embryo by having it transferred to her womb is married, and makes this choice with her husband's consent, this seems to be preferable to having an unmarried woman making this choice. It is preferable because the married couple can adopt the child, and the wife would choose to have the child transferred to her womb as a first step in adopting the child, whose first "home" would be her womb, and who would then be given the home by the husband and wife together it needs and merits. I believe that it is preferable for a married woman to have the embryo transferred to her womb from the freezer because after its birth she and her husband can formally adopt it. It would not then be necessary for the unmarried woman—who had acted uprightly, as we have seen, in having it transferred to her womb from the freezer as a means for protecting its life—to have it adopted by a married woman and her husband.

Some Caveats

There is, unfortunately, the danger that "rescuing" frozen embryos in this way *could* lead to commercializing the procedure. In fact, one business firm in the U.S., Creating Families, Inc., Denver, Colorado, proposes that it locate and distribute abandoned frozen embryos to couples anxious to have a child. Although the company is careful to claim that it does not accept "fees" for its work in "creating families," it solicits "donations" for its services, and it would be unrealistic to think that the profit motive is not operative here. Similarly, the Columbia Presbyterian Medical Center in New York provides an already frozen embryo to couples for $2,750 as opposed to the $16,000 needed to cover the entire IVF-ET procedure.[11] Commercial or quasi-

---

[11]On the Columbia Hospital "offer," see Brian Caulfield, "Souls on Ice: With Frozen Embryo Technology Life's Sanctity Is Lost," *National Catholic Register* 74.1 (January 4–10, 1998): 15.

commercial ventures treating frozen embryos as commodities are obviously immoral, and surely individuals and/or couples can resort to transferring frozen embryos to the women's wombs (either married women, single women, or women seeking to act as "surrogates" for others) as a means of satisfying their desire to have a child. The *object* freely chosen by such individuals and/or couples is *not* to have a frozen embryo transferred into a woman's womb as a means of protecting its life but is rather effecting such a transfer as a means to satisfy their own desires. This is, of course, a different moral object, and one that is not compatible with the love and respect due to the life of the unborn frozen embryo who is regarded as a commodity.

There is also the danger, voiced by Bishop Elio Sgreccia, past vice-president of the Pontifical Academy for Life, that "rescuing" frozen embryos *could* serve to legitimize and encourage the production of embryos in vitro and freezing them.[12]

But this horrible abuse of frozen embryos, although a serious danger and one, too, that may not be preventable, does not make the morally good choice of having a frozen embryo transferred from its concentration can as a means of protecting its life, into something evil, because different moral objects are involved.

Reply to Objections Raised by Reputable Catholic Philosophers and Theologians

Here I will respond to the arguments advanced by Msgr. William B. Smith, professor of moral theology at St. Joseph's Seminary, Dunwoodie, N.Y., and Mary Geach, a British philosopher, wife, and mother.

Smith thinks that "rescuing" a frozen embryo in the way defended above is intrinsically immoral and that there are no morally legitimate ways of saving the lives of these abandoned tiny human persons. He contends that his view is supported by a specific passage in *Donum vitae,* by that document's teaching on surrogate motherhood, and on the "principled conclusion" of the document.

---

[12]Bishop Sgreccia is quoted to this effect by Marco Tossati, "Chiesa divisa sugli embrioni," *La Stampa* (July 22, 1966). See also Mauro Cozzoli, "The Human Embryo: Ethical and Normative Aspects," in *The Identity and Statute* [sic] *of Human Embryo: Proceedings of the Third Assembly of the Pontifical Academy for Life,* eds. Juan de Dios Vial Correa and Elio Sgreccia (Vatican City: Libreria Editrice Vaticana, 1999), 295.

The specific passage on which he relies is the following: "in consequence of the fact that they have been produced in vitro, those embryos which are not transferred into the body of the mother and are called 'spare' are exposed to an absurd fate, with no possibility of their being offered safe means of survival which *can be licitly pursued*."[13] Commenting, Smith then says: "No safe means that *can licitly be pursued!* Perhaps the CDF did not intend to address this particular case [that of a woman who proposed to "adopt" a frozen embryo pre-natally and have it transferred to her womb], but I read here a first-principled insight indicating that this volunteer 'rescue' is *not* a licit option."[14]

Smith also contends that the document's teaching on surrogate motherhood excludes this option. While granting some differences between the aims of the "surrogacy" usually associated with in vitro fertilization and the aims of the woman seeking to "rescue" a frozen embryo, he nonetheless argues that the "foundational reasons for rejecting 'surrogacy' as illicit also apply to this project," namely, "a failure to meet the obligations of maternal love, conjugal fidelity, and responsible parenthood."[15]

The "principled conclusion" of *Donum vitae* to which Smith appeals is its teaching on the inseparable bond uniting the unitive and procreative meanings of the conjugal act and between the goods of marriage, along with the unity and dignity of the human person which requires that "the procreation of a human person be brought about as the fruit of the conjugal act specific to the love between the spouses."[16] Since the projected "rescue" is not procreation of this kind, Smith concludes that it cannot be morally licit.[17]

Both Grisez and Geoffrey Surtees argue, correctly in my judgment, that Smith has taken the passage in *Donum vitae* to which he appeals out of context. It occurs in a section where concern focuses on using embryos produced in vitro as subjects

---

[13]*Donum vitae*, I, 5, cited by Smith in his essay, "Rescue the Frozen?" *Homiletic and Pastoral Review* 96.1 (October, 1995): 72. Smith adds the emphasis to this passage.

[14]Ibid.

[15]Ibid.

[16]Ibid, 74, citing *Donum vitae*, II, 4.

[17]Ibid. I believe that here I have accurately summarize Smith's view, although it is not too clearly put, in my opinion.

of experimental research. Thus Grisez concludes that the "sentence Smith quotes should not be understood as referring to the action of a rescuer who has in no way participated in the wrongs that have brought the embryonic persons to be and left them to their absurd fate, but to the options available to those wrongly involved in IVF."[18]

Smith's second line of defense—that rescuing frozen embryos by having them transferred to the womb of a woman who volunteers to rescue them is excluded by *Donum vitae's* condemnation of surrogate motherhood—is also answered effectively by Grisez and Surtees. Grisez notes that Smith "ignores the fact that bearing [a child] *on another's behalf* is part of the very definition of surrogacy."[19] To this Smith undoubtedly would reply, "Yes, that is true. But the 'foundational reason' for rejecting surrogacy is nonetheless still operative, namely, a failure to meet the obligations of maternal love, conjugal fidelity, and responsible parenthood." In his essay Surtees emphasized that the *object* freely chosen by the woman who saves the frozen embryo by adopting it prenatally[20] is completely different from the object freely chosen by those who resort to making a child through in vitro fertilization or who freely choose to act as surrogate mothers, i.e., bear a child for some other person's benefit.[21] Smith, responding to Surtees' argument, insisted that, although the intended *end* is good (to rescue the baby) the *means chosen* is not licit because the means chosen is to "become a nine-month-surrogate" in order to adopt the child.

Smith's claim, I believe, is inaccurate. The woman is definitely *not choosing* to nurture the child in her womb for the benefit of someone other than the child—and this is what the surrogate does. Although Surtees may have inaccurately identified

---

[18]Grisez, *Difficult Moral Questions*, 242, note 188. See also Geoffrey Surtees, "Adoption of a Frozen Embryo," *Homiletic and Pastoral Review* 96 (August–September 1996): 8–9.

[19]Grisez, *Difficult Moral Questions*, 241, note 186.

[20]Surtees, "Adoption of a Frozen Embryo," holds that the *object* chosen is "to adopt a frozen embryo prenatally." I believe that Grisez has more clearly identified this object, namely, "to have the frozen embryo transferred in a woman's womb," but Surtees sees, as the first step in adopting the orphaned embryo, its transferal from freezer to the woman's womb. Hence his view is very close to that of Grisez and differs from Smith precisely because Surtees holds that the object morally specifying the act is neither to generate life outside the marital act, nor to act as a surrogate, nor to violate the marriage.

[21]Surtees, "Adoption of a Frozen Embryo," 12–13.

the object chosen as "adoption of the frozen embryo" rather than precisely as the "transfer of the orphaned embryo from its freezer to the womb of a mother," his identification of the object is closer to the truth than is Smith's, who sees the chosen means precisely as surrogacy. Were he to put himself in the perspective of the acting person, i.e., the woman who volunteers to have the embryo put in her womb, he would understand the object more clearly, for good pro-life women who have volunteered to rescue embryos in this way *rightly* reject surrogacy as the object of their choice. And, as John Paul II emphasized in the passage cited earlier, we *must* put ourselves in the perspective of the acting person if we are properly to identify the object freely chosen.

The "principled conclusion" of *Donum vitae* to which Smith appeals is a true principled conclusion, namely, that "the procreation of a human person be brought about as the fruit of the conjugal act specific to the love between spouses." But the human act freely chosen by the woman who volunteers to have the frozen embryo transferred to her womb as a means of saving its life is surely not "to procreate a human person outside the conjugal act." A human person in this instance has *already* been generated outside the marital act—and those who choose to produce the child in this way surely violate the "principled conclusion" of *Donum vitae*. But the rescuing woman does not, nor does she choose to violate conjugal unity and responsible parenthood. She rather freely assumes responsibility to save and protect a human life by giving the support it needs at this stage of its development.

Mary Geach regards efforts to rescue frozen embryos by having them placed in the womb of a woman as intrinsically evil. A woman doing this is engaging in an intrinsically evil act because she is violating her reproductive integrity. Such integrity requires that a woman become pregnant voluntarily *only* through normal marital relations. But the woman seeking to rescue the frozen embryo is making her womb available to strangers and allowing herself to be made pregnant by means of a technical act of impregnation. By doing so she shares in the evil of in vitro fertilization and ruins her reproductive integrity.[22]

Geach says, quite correctly, that when a married woman engages in the conjugal act, a unitive act of the procreative kind, her providing the means to an impregnating intromis-

---

[22]Mary Geach, "Are there any circumstances in which it would be morally admirable for a woman to seek to have an orphan embryo implanted in her womb?" in *Issues for a Catholic Bioethic*, 341–346.

sion "is a vital part of the self-giving involved in the woman's part of the marriage act," and this "self-giving is not just a self-giving to the possible child, but to the father, since it would be his child that she would be bearing." She then claims that the woman's act of admitting a frozen embryo into her womb and thus allowing herself to become pregnant "is profoundly like such an act of generation." Indeed, she continues, "this similarity is more profound that the similarity of a woman's perverse sexual act [masturbation] to the female act of generation," because a "complete marriage act is *always* an act of admission which is of a kind to make one pregnant.... To separate it [the act of admission of a kind to make one pregnant] from the other parts of the marriage act would be to destroy in oneself that same reproductive integrity which is destroyed by unchaste actions." It thus follows that if it seriously wrong for a woman to excite within herself through masturbation sensations properly associated with and accompanying the self-giving involved in the conjugal act, then "how much worse it must be to isolate the spiritual component of the marriage act, the giving up of the body to the impregnator, dissociating oneself from the parents of the child, and substituting for the relation to the father a mere arrangement with a technician."[23]

If Geach's analysis of *what* the rescuing woman is in truth freely choosing to do were correct, then I would be in perfect agreement with her. The moral object specifiying her freely chosen human act would be intrinsically evil, "giving up her body, not to her husband, but to a technician," "dissociating herself from the parents of the child," "substituting for the relation to the father a mere arrangement with a technician." According to Geach, these seem to be the "objects" specifying the woman's choice. She is allowing herself to become pregnant outside of marriage.

In many ways I believe that the arguments set forth in the second part of this essay have laid the basis for replying to Geach. There I argued that in rescuing an abandoned embryo by having it transferred from the freezer to her womb, the moral object specifying the woman's choice is precisely to "transfer the embryo from freezer to womb as means of protecting its life." In setting forth the reasons supporting this claim I cited abundantly from the essay by Helen Watt, an essay written explicitly as a rebuttal to Geach's argument. As noted above, Watt clearly

---

[23]Ibid., 344–345.

distinguishes the "intentions" operative in allowing a child "to be created in one's womb" outside of marriage and in allowing an already existing child (whose life is in peril) "to be put in one's womb."

Geach, in a reply to earlier similar criticism of her position[24]—charging that she erroneously identifies the object freely chosen—maintains that this criticism is arbitrary and that I (and Watt presumably) am *redescribing* the object in question in much the same way as anyone wishing to kill his old uncle to get money could say that

> his *not* intrinsically evil object is that he inherit money under his uncle's will. But he wishes to kill in order to do this, and he cannot choose the description 'act liable to lead to my inheriting money' and say that that specifies his action morally.... One is responsible for one's actions under all descriptions under which one freely chooses them. In the case at hand, the woman freely chooses to admit an insertion of a kind to make her pregnant without generating.... It is under this description that her action is unacceptable, since it follows from this description that she is performing an act which is centrally and importantly like the marriage act, but which is not the marriage act or a part of it.[25]

I obviously agree with Geach about the man who kills his uncle and attempts to justify it in the way he does. He *is* redescribing his action *in terms of its hoped-for benefits,* a common enough way of rationalizing evil deeds. His *end* is to inherit money and the *means chosen* (=object of the act) is to kill his uncle.

But such redescribing is *not* being done in the rescuing of frozen embryos. In saying that the woman rescuing the embryo is *transferring the frozen embryo to her womb,* one is in no way redescribing what she is doing here and now in terms of hoped-for future benefits. Nor is she choosing to get pregnant outside of marriage. This is not the *object* freely chosen. She foresees that she will become pregnant after the embryo implants in her womb, but this is not precisely what she is choosing to do here and now, she surely is not choosing to become pregnant

---

[24]Geach sent me two emails, one February 17, 2000, and another February 20, 2000 to answer criticism of her work voiced by Watt and me.

[25]Geach email to me, February 20, 2000, printed in my *Catholic Bioethics and the Gift of Human Life* (Huntington, IN: Our Sunday Visitor, Inc., 2000), 100.

by giving herself sexually to one who is not her husband. The woman who rescues the embryo is responsible for allowing herself to become pregnant as a consequence of having the embryo put into her womb, but she did not make the immoral choice to get pregnant outside of marriage. Geach seems to think that she did.

In sum, in volunteering to have an orphaned frozen embryo transferred into her womb as a means of saving the life of this child, a woman's object of choice, which is the primary determinant of the morality of her act, is not to engage in an act intrinsically evil.

RESCUING FROZEN EMBRYOS

MARY C. GEACH

The rescue of frozen embryos at the present time requires that women allow themselves to be made pregnant with children[1] not their own. Whether artificial wombs could be invented which might enable us to rear embryos without providing mothers for them I do not know, though I very much doubt it. The practical question before us is this: can it be right for a woman to allow herself to be made pregnant with a child which is not her own?

Dr. May might wish to deny the relevance of this question.[2] He maintains that a woman who allows a frozen embryo to be transferred to her womb is not intentionally allowing herself to be made pregnant outside marriage; that the pregnancy

---

[1] I assume throughout this paper that an embryo is a human person. I do not need to justify this assumption, as it would make my case even stronger were it true that an embryo was not possessed of a human soul.

[2] See William E. May, "The Morality of 'Rescuing' Frozen Embryos," in the previous chapter, 201–215.

is a foreseen but unintended consequence of her action. He rightly distinguishes between consequences which are intended and ones which are foreseen. All kinds of descriptions of a human action, not only as to consequences but also as to circumstances, etc., may or may not be descriptions under which the action is intended, and whether it is intended under some description does not depend simply on our state of awareness. As I walk along, I disturb the air, and I know that I disturb the air, but this does not mean that I intend to disturb it. So how do we distinguish the descriptions under which an action is intended from those under which it is merely averted to?

We make this distinction by asking "Why?" Why are we doing what we are doing? What for? Why is the professor failing the degree candidate? Not, he says, in order to deprive him of the degree. But suppose that his reason for failing the degree candidate is that he thinks that too many people are getting degrees in his subject and wishes to make sure that there are fewer. He cannot *then* say that the student's not getting his degree in his subject is a foreseen but unintended consequence of flunking him. The professor thinks of the flunking as a desirable action precisely because it will mean that there is one fewer degreed person in his subject. Depriving the candidate of the degree is then a means to the professor's end in flunking the man, and cannot be regarded as unintentional.

Let us then return to the woman who has a frozen embryo put inside her. Why is she allowing this embryo to be placed in her womb? Because she wishes to save his life and to allow him to develop normally. The only way to save his life, and allow him to develop normally, is to place him in a womb, in which he can develop as embryos do in the wombs of their mothers. So she allows the embryo to be inserted. But to have a child in one's womb, developing as embryos do develop in the wombs of their mothers, *is to be pregnant with that child.* Thus the woman's being pregnant with the child is a means to her end (of saving his life) and is an intended consequence of her action. We intend not only the ultimate goals of our actions, but also the means by which these goals are achieved. In this case, the means is pregnancy.

So now let us return to the question "Can it be right for a woman to allow herself to be made pregnant with a child not her own?" I hold that the action of a woman who allows this is intrinsically evil. If it is not, it is hard to see why surrogate motherhood should not be allowed. You might say, surrogate motherhood involves perverse action on the part of the child's genetic parents, but it does not necessarily do so. Some women, who are quite capable of conceiving normally, have wombs in which it is impossible for the child to come to term. Some embryos cannot implant in their mothers. Now there exists a technique called "embryo flushing" by which doctors can remove an unimplanted embryo from the womb. If it is allowable for a woman to allow herself to be made pregnant with a child not her own, why should a woman with this problem about bearing her children not make use of another woman's womb for the purpose of bringing her own child to maturity? Professor Grisez has compared the woman who bears an orphan embryo in her womb with a woman who suckles an orphan, feeding him with her own milk.[3] He does not think, I would guess, that the practice of having a wet nurse was a good one when ladies did it to avoid the trouble of having to do the job themselves, but a woman who was dry, or too busy or ill to feed her own child, might get another woman to be wet nurse, and this would be work for which she could quite reasonably be paid. Similarly, a woman might bear another woman's child because the other woman could not bring hers to term, and might quite reasonably be paid for this if (as must be the case if frozen embryos may be rescued) there is nothing intrinsically evil about allowing oneself to be made pregnant with another woman's child. Professors Grisez and May distinguish between embryo rescue and surrogacy, but if embryo rescue is permissible, there is no clear reason for forbidding surrogacy. If bearing another woman's child is like breastfeeding another woman's child, then one can do the one thing on another woman's behalf just as one can do the other. Thus the womb of the woman, from being a private space where only her husband's seed may be sown, can become a public resource, a facility to be paid for.

Dr. Helen Watt, who does not say with Professor May that the pregnancy of the embryo-rescuing woman is an unintended effect of her action in allowing the embryo to be implanted, could reply to this point of mine by saying that a woman who has a child implanted in her becomes a mother, and that one should not undertake motherhood on a temporary basis.[4] For the same

---

[3]Germain Grisez, *Difficult Moral Questions*, vol. 3, *The Way of the Lord Jesus* (Quincy, IL: Franciscan Press 1997), 241.

[4]Helen Watt, "Are there any circumstances in which it would be morally admirable for a woman to seek to have an orphan embryo implanted in her womb?" in *Issues for a Catholic Bioethic*, ed. Luke Gormally (London: The Linacre Centre, 1999), 347–352 at 347.

reason, she thinks that a woman who does not intend to bring children to term is morally obliged not to become pregnant intentionally. (A harsh view, implying that a woman who knows herself to have a womb incapable of bringing a child to term is forbidden to marry.) She sees an important difference between being a wet nurse and bearing a child in the womb. A child, once implanted, cannot be moved from womb to womb; a child can be moved from nurse to nurse. So becoming pregnant with a child involves a commitment which one should not undertake unless one is prepared to carry it through.

I think that Dr. Watt's reason for distinguishing between the embryo rescuer and the wet nurse is a rather weak and poor one, since there is usually only one wet nurse, and since it is good, and not bad, that, in this period of a child's life, there should be just one woman who fulfils this nurturing role, and who thus is the maternal figure for the child. (Unless indeed the real mother might be able to suckle her child part-time, which perhaps was what Dr. Watt was imagining.) In any case, there is, I believe, a distinction to be made, and it is absolutely crucial that one should grasp this distinction. If a woman voluntarily has an embryo placed in her womb she is performing *an act of admission whereby she allows an intromission of impregnating kind to be made into her.* This act is centrally like the marriage act, as performed by a woman. Nothing which a wet-nurse does shares such a central description with the female act.

Professor Grisez thinks that embryo rescue involves no assault upon the good of marriage, because it is not sexual. He thinks like this because he has not reflected adequately on the woman's part in the marriage act. In this act, it is necessary for the man to have the motions of the flesh which belong to the act. If he does not, the act has not taken place, no matter what may be the sensations, etc., of the woman. The reverse is the case with the woman. If she does not have any of the sensations and motions of the flesh which are associated with the marriage act, that is sad for her but does not mean that she is not performing the act. What is essential is that she performs an act of admission, allowing an intromission which is of a kind to make her pregnant.

Some women have married with only a vague idea of what the marriage act consisted in—this did not, I think, invalidate their marriages, so long as the woman understood that the effect of what the man did was to put into her something which was of a kind to make her pregnant. But if the woman did not know about *that*, about getting pregnant, then clearly it was no marriage at all.

Similarly, a woman might be rather vague as to what took place inside. Does the man sow a seed, already having the potential to grow into a man? Does he just activate a seed already in her? Again, this sort of ignorance would not invalidate either the marriage or the marriage act. What is essential is that the woman know that the intromission by the man is of a kind to make her pregnant.

So if a woman admits an intromission which is of a kind to impregnate her, she is performing an act which shares a central and essential description with the marriage act. This act of admission misses out some other central and essential descriptions: it is not of generative kind, for instance. This means that the woman's act is a highly defective version of the marriage act. That her intention is significantly different from the intention of a woman performing the marriage act (a point made by Dr. Watt) is an essential part of my objection to what she does.

Now, it was from reading, some years ago, what Professor May had written on these topics, that I first came to see that the position of the Church on this subject is one which requires of us something which may be called reproductive integrity.[5] Reproductive integrity means not taking parts and aspects of the marriage act and setting them up on their own, which we do if we masturbate or go in for perverse and unnatural methods of reproduction, or if we have a sexual relationship outside marriage. If we took part of a machine and used it on its own as a tool, without respect for its nature as a part, we would reduce its capacity to function as a part. Even a machine should not be used as a collection of items, and still less should an organic whole like marriage. If marriages are always going wrong, it is because this is done, and one part or aspect of the marriage act is treated as a thing on its own. One part or aspect of the marriage act is the woman's act of admission, whereby she allows an intromission which is of a kind to make her pregnant.

In this context we should reflect on the Annunciation, which in these matters is the exception which proves the rule. The rule in question is that a woman should not allow herself

---

[5]See the discussion of the contrast between "integralist" and "separatist" understandings of human sexuality in William E. May, *Sex, Marriage and Chastity* (Chicago, IL: Franciscan Herald Press, 1981).

to be impregnated except through an act which on her part is the natural and central expression of a profound, permanent, and sexually exclusive relationship with the father of her child, who by his act of impregnation generates his own child within her. In this quite exceptional case, where the impregnator does not perform a sexual act, this rule still holds. God is not exploiting Mary, because the rule still holds. But when a woman allows herself to be impregnated, and her act expresses no such relationship with the father, one part of her, the right to say yes or no to the impregnating intromission, is treated as a thing on its own, not forming part of the nuptial wholeness, the marriage act.

In a more recent contribution to the debate,[6] Dr. Watt says that I take artificial insemination and "embryo rescue" to be objectionable because they are *nonsexual* means of becoming pregnant. But what is objectionable about these is that the woman's act is not the central expression of a relationship with the begetter. The central expression of Mary's relationship with the Father is "be it done unto me according to thy word"; this can only be the proper and central expression of a relationship with God, who only by his word alone can make things come to be.

We are taught that Mary is ever-virgin, because the relation one should have with the father of one's child does not allow, in his lifetime, a self-giving of the same kind to anyone else, and God lives for ever. Mary, in allowing herself to be impregnated, has expressed a permanent, sexually-exclusive relation to the father of the child. However, she is not performing an act of admission allowing a spatial intromission, so she is not performing a version of the marriage act, though in her relation to God the same reason for sexual exclusiveness exists (*par excellence*) as in the case of the marriage relationship. Only by treating the woman's body as a sacred vessel, made by her act of openness to the father into a container of their life's continuance, is the fact respected that her living body (the body not being alive unless informed by the soul) is herself. For a woman, to give her body is to express a relationship in which her life is exclusively bound to another's. If no such relationship exists, the capacity of her acts of self-giving for expressing such a relationship is damaged.

---

[6]Helen Watt, "A Brief Defense of Frozen Embryo Adoption," *National Catholic Bioethics Quarterly* 1.2 (Summer 2001): 151–154.

If self-giving to one's husband results in the conception of a child, that event itself expresses the bond between husband and wife. If, however, a woman's openness to an act of intromission results in her being pregnant with the child of someone other than her husband, what has occurred is contrary to the bond between husband and wife. And the resultant alienation will be manifested, for example, in the fact that, at least in the earliest stages of pregnancy, the husband, in order not to disrupt implantation, will be excluded from marital intercourse with his wife.

Dr. Watt[7] objects to my seeing the woman's act of admission in embryo transfer as being like her act of admission in the marital act, saying "Why is embryo transfer not sufficiently *removed* from the generative act to make it, not a perversion of that act, but something else entirely?" She here speaks of "embryo transfer" as the act to be compared to the marital act. Perhaps the act whereby the embryo is transferred is not like the male act. But the description "act of admission allowing an intromission of impregnating kind" is very like a description of the female act. Dr. Watt's objection fails to attend to my actual description of the woman's part in embryo transfer, which is a pity as it is precisely this description of the woman's act in allowing embryo transfer that makes me call her act a distorted version of the wife's part of the marriage act.

Dr. Watt has objected that to be made pregnant may mean to generate a child, or it may mean to have a child implant in the womb, and that the direct effect of intercourse is generation and not implantation. However to become pregnant is *to begin to have a child within*.[8] Since embryo rescue involves actually putting the child in the womb, the effect of the intromission admitted by the woman is directly to make her pregnant, and the intromission is in itself of impregnating kind. This also gives us reason to say that the woman's intention is to allow an intromission of impregnating kind. To show this I argue as follows: to put something into something is to bring it about that the thing put in is in there. So (if the thing was not there before) to put something into something is to make the one thing to begin to be in the other. Therefore a woman who

---

[7] Ibid., 152.

[8] And so abortion, understood as the termination of a pregnancy, is rightly said to take place when the death of an embryo is deliberately caused by preventing implantation.

admits an embryo into her womb, when that embryo has not been in her womb before, is allowing it to begin to be inside her. That is, she is allowing something to be put into her the putting in of which is making her pregnant: this follows logically. Now, we cannot be said not to intend what follows logically from what we do intend. Even Professor May would presumably agree that, in allowing the embryo to be "transferred" to her womb, she is allowing to be put inside her an embryo which has not been in her, and that this is an intentional act. She is therefore performing an intentional act of allowing a putting into her of something the putting in of which is making her pregnant. But this is to perform intentionally an act of admission whereby she allows an intromission which is (in itself) of impregnating kind, being of a kind to make her pregnant and not having been in her before. (She intends it as not having been in her before because she has chosen the embryo as having parents who have abandoned it.)

Dr. Watt has raised two serious objections to what I have to say, or rather to what I said in a paper I gave on this topic in 1997, which Professor May's accompanying paper also criticizes.[9] What she has said has caused me to tighten up my description of the woman's act of admission, about which I had said in some places that it allowed something to be done to her which was of a kind to make her pregnant and objected to it on that ground.

I will deal first with an objection which she has raised in conversation with me. She said, suppose a woman had already had intercourse with her husband and there were sperms in her fallopian tube. And suppose the ovum were then moved by a doctor past a blockage in the tube which was preventing her conceiving. This movement of the ovum is called lower tubal ovum transfer, and is rightly regarded as licit by orthodox moral theologians. Well, says Dr. Watt, is this not an intromission of a kind liable to make the woman pregnant?

I reply that the human act of making an intromission of an ovum under the circumstances described is an act of a kind to make the woman pregnant, but that the circumstances (i.e. that the tube is full of sperms) have to be mentioned as part of the description of the act. So the intentional act which is of a

---

[9]Mary Geach, 'Are there any circumstances in which it would be morally admirable for a woman to seek to have an orphan embryo implanted in her womb?" in *Issues for a Catholic Bioethic*, ed. Gormally, 341–346.

kind to make the woman pregnant is not "putting in the ovum" but "putting in the ovum when the tube is full of sperms." This act is of a kind to make the woman pregnant because it is ancillary to the marriage act, and being only ancillary, is permissible. However, the intromission of the unfertilized ovum is not in itself considered simply as an intromission (and *not* as a human act), one which is of a kind to make the woman pregnant. Similarly, one might be given an injection which made one ovulate shortly after intercourse, and in this case also, though the human act of the one giving the injection is an act of performing an intromission which is of a kind to make one pregnant in the circumstances, the circumstances again have to be mentioned as part of the act; the intromission in itself is not of a kind to make one pregnant, is not an impregnating intromission.

By contrast, an intromission of sperm, whether by natural or unnatural means, is an intromission of a kind to make one pregnant, since it is by the intromission of sperm that women normally become pregnant and since it is the function of sperm to enter the woman's womb from outside and to fertilize her ova, thereby making it the case that she is pregnant. The intromission of a fertilized ovum is also an intromission which is of a kind to make the woman pregnant if she has not already become so; in this case, because to begin to have a child inside *is* to become pregnant.

When the ovum is moved to make it accessible to sperms already present in the fallopian tube, the marriage act is thereby allowed to take effect, so the woman's act of allowing this movement obtains its character as generative in being ancillary, or subsidiary, to her marriage act. An act subsidiary to another act has the same end as that other act, but is not therefore a *version* of the type of act specified by that end.

Dr. Watt has also raised the possibility that it might be medically advisable to remove the child and then replace it.[10] When she first raised this possibility, I did not take it seriously, but perhaps a surgeon operating on a blocked tube might find that the ovum was fertilized already. Sperms might, I suppose, be able to get past a blockage which ova could not pass. Also, as Dr. Watt has pointed out, some surgeons have *tried* to reimplant embryos when they have removed the tube with a baby in it. So, she imagines, one might be offered the possibility of an

---

[10]Watt, "Are there any circumstances," 350.

attempted reimplantation. Under these circumstances, a woman who did not know that the attempt was futile might agree (without doing wrong) to having the child re-placed. (I do not know if it would be futile, so I would agree to such a procedure myself.)

To this point of Dr. Watt's I say that, as the intromission admitted by the woman does not initiate her pregnancy, but only resumes it after an interruption, her allowing the intromission of the embryo does not have the similarity to the act of intercourse which makes it a distorted version of that act. Permitting the intromission of a child having its origin in you is unlike permitting the intromission of a child not your own, as accepting payment is different from accepting a gift; your pregnancy would not be caused by the moving of the embryo; the intromission would thus not be of impregnating kind, but would be adjusting a maladjusted pregnancy. Part of the concept of impregnation is that the intromission is of something whose origin, whose power to make you pregnant, comes from another, not from yourself.

Another objection by Dr. Watt is that a woman who has conceived by IVF and has embryos in store is obliged to bear them[11]; but what we are obliged to do, we are necessarily permitted to do; but we are never permitted to do what is evil in itself;[12] therefore it must be alright to be made pregnant by having embryos put inside one; therefore it must be alright to do this for orphaned embryos to save their lives.

The doubtful premise in all this is the second one: that what we are obliged to do we are necessarily permitted to do. A woman who leaves her children in a freezer to die is clearly doing them wrong, by depriving them of the nurture which she owes to them. But we cannot always pay our debts; we cannot

---

[11]Dr. Watt asserts this obligation in ibid., 348. In the remainder of the paragraph in my text, I make clear the way Dr. Watt has completed her argument in conversation in order to draw the conclusion she does about the acceptability of adopting orphan embryos.

[12]A point which can be made against this argument is that sometimes our conscience commands us to do what is evil in itself. It follows from this that either we are not always obliged to follow our conscience (an unacceptable position), or we can be permitted to do what is evil in itself, or (as Aquinas concludes) we may be, through our own fault, obliged to perform impermissible actions. For Aquinas on moral perplexity *secundum quid* see *Summa theologiae* I-II, Q. 19.6, reply 3; II-II, Q. 62.2, obj. 2; III, Q. 64.6, reply 3; *De veritate* 17.4, reply 8.

always do for others what they have a right to. Physically, this can obviously be the case, and if the reason for its being physically impossible is that we have made it so, the physical impossibility does not lessen the wrong done to the other. But just as we can make it physically impossible to pay a debt, so we can make it morally impossible also. An argument to show that a kind of action must be right because people have in some wicked way made it impossible to pay what they owe without performing that action is no argument at all. Wickedness leads to perplexity. What the perplexed are to do is another question. Can a perplexed person form a firm purpose of amendment? I do not know. Anyway, the fact that Jones, through his own fault, can only pay his just debts by handing over money which he is holding in trust does not show that handing over such money is not stealing it. So, if it is owed to the embryo that he should be nourished in his mother's womb, this does not necessarily show that it is right to insert it there: but it might be argued that this case of inserting an embryo is not an impregnating intromission in the relevant sense, since the origin of such an embryo is in the mother, so that the otherness of impregnation is not there.

Dr. Watt says why, if the origin of the embryo is in the mother, does this make a difference to her act in allowing the intromission? Why should the one be perverse, and the other not perverse? Why should one be seen as having a central and significant resemblance to the marriage act, which the other does not have?

Now, in the case of a mother's own IVF child, I am uncertain; but in the rather unlikely event of an offer to restore an interrupted pregnancy, it is clear enough that there is a great psychological difference here. One would naturally express by saying, not that one had been made pregnant by the doctor's operation, but rather that the doctor had restored one's pregnancy. This is because the relationship into which one enters in becoming pregnant is one which persists even if the pregnancy is interrupted. As Dr. Watt puts it, one becomes a mother by being made pregnant with an orphan embryo, but if one has one's child put back (supposing this to happen) one is a mother already.

So, Dr. Watt might say, what you are objecting to is not an intromission which is of a kind to make you pregnant, but one which is of a kind to make you a mother. And since it is alright to become a mother by adopting a child, why is it not alright to become a mother by adopting an embryo? (Or perhaps: why is it not alright to adopt an embryo, and then have it inserted, since one has by adoption become its mother, and the proper place for an embryo is in its mother's womb?)

There is a difference between becoming a mother by adoption and becoming a mother by implantation. If one does the former, one's own action is what makes one the mother, but if one does the latter, it is the action of the inserter that makes one the mother of the child inserted. One's own action is to admit the insertion in the latter case; in the former, one's own formal undertaking is precisely what makes one a parent. The adoption relation is a legal one, and carries legal obligations, and moral obligations, in respect of this legal relation. The relation to a child who is in one's womb is natural or quasi-natural, and it is in respect of this natural or quasi-natural relation that one has such maternal obligations as may accrue. The maternal obligation involved in adoption does not oblige one to enter into a quasi-natural biological relationship, unless indeed it makes sense to say that one can become a mother by adoption of an embryo, and thereby acquire such an obligation. But one cannot become a mother by adoption of an embryo: to adopt a child is to acquire the normal responsibilities of a parent, but there are no normal responsibilities of a parent to a child still in the embryonic phase, save to act in such a way as to preserve the biological tie. Dr. Watt would have it that one can be obliged to bring this biological tie into being.

However, my objection still remains, that by allowing the impregnating insertion one is performing an act which is like the marriage act in its spiritual essence, an essence which requires that the father of the child be playing his part. If one does without him, one is exploiting oneself: for to exploit someone is to use that person as a means in a way which ignores and cuts across that person's ends and goods. The end and the good of one's womb is in one's relation to the father of the child, as well as to the child. But if one has something done to one's own child, one's relation to whom has already become a part of the natural good for which the womb exists, one is not cutting across or ignoring the end for which the womb and the use of the womb exist, nor that for the sake of which the woman has the ability to allow, and the right to deny, access of an impregnating kind to this part of her genital tract: and that is the good of marriage.

A particular sort of self-sacrifice, whereby a woman surrenders her whole body for the sake of her husband, and of her husband's children, is a part of uncorrupted female sexuality.

In the marriage act, this particular feminine sexuality comes into play. Corrupted, it makes women ready to do the most dreadful things and submit to the foulest indignities. I pray that no one in a position of pastoral responsibility in the Church becomes a party to corrupting, by exploiting, this great force in women, and leads them to submit to impregnation, without this being an expression of a preexisting and permanent commitment, of a sexually exclusive kind, to the one who generates the child within them.

Supplementary Note

Whether the woman's act is accompanied by the feelings and motions of the flesh appropriate to the essentially feminine surrender of the flesh, her act is such a surrender, a surrender to an act of intromission of impregnating kind. Professor May says that a woman's intention in allowing such an intromission is not "intending to give herself sexually to one who is not her husband," but her act (whether or not it is accompanied by erotic feelings) remains like enough to what she does in the marriage act to be a version of what she does in that act. Professor May's implausible contention that the woman does not choose to get pregnant "outside of marriage" when she admits the embryo testifies to his sense that such a choice—the choice to admit an intromission of impregnating kind—is an immoral one when it is not a marriage act. What I say at the beginning of this paper and on pages 225 and 226 adequately replies to Professor May's contentions about intention.

Dr. Watt has recently pointed out[13] that the intention to have an embryo generated inside one's body is significantly different from the intention to have an embryo transferred to one's body, and says "Is it therefore so obvious that embryo transfer, as opposed to artificial insemination, is a perverse variation on the generative act? Why is not embryo transfer sufficiently *removed* from the generative act to make it, not a perversion of that act, but something else entirely?"

To this I reply that I did not call embryo transfer a perverse variation on the generative act. I typified as such a perverse variation the *woman's* act of admission whereby she allows an intromission of impregnating kind. People who write about this question fail to recognize, or shy away from speaking about, the question of *the nature of the female marriage act.* Dr. Watt's speak-

---

[13]Watt, "A Brief Defense," 152.

ing of "embryo transfer" exemplifies this. We know what men do in the marriage act: What does the woman do? No satisfactory answer to my argument can by given by anyone who does not answer this question, nor by anyone who fails to see that the description under which I object to the woman's act in embryo rescue is one which is a) an accurate description of what the woman does, and b) a description of her act which, though vague about the material aspects, is so like a description of what the woman does in the marriage act as to be downright indecent. The indecency of saying that the woman performs "an act of admission whereby she allows an intromission of impregnating kind" does not itself show that what she does is wrong, but it does show that what she does is like the marriage act.

Dr. Watt's failure even to quote the description under which I object to the woman's act in embryo rescue resembles Dr. May's unsustainable contentions about what is intended in embryo rescue. Each author shows, in their own way, that they have the sense, which anyone must have, that the description "an act of admission allowing an intromission of impregnating kind" is an unacceptable description for a non-generative kind of act which expresses no profound or permanent relation to the father of the child.

RAPE PROTOCOLS AND ABORTIFACIENT ACTIONS OF ORAL CONTRACEPTIVES

RALPH P. MIECH, M.D.

Secular Manipulation of Semantics

Secular ideology uses a combination of cultural, social, educational, and legal programs to promote the acceptance of secular mores in the field of human reproduction. To achieve this end, secular programs utilize the prime tool of changing or creating definitions of key medical terms. The purpose of these secular programs is to mislead the public and undermine traditional moral teachings in the area of human reproduction.

The prime example of secular manipulation of semantics was the change in the definition of what is a drug. Prior to the 1950s a drug was defined as any chemical substance used in the diagnosis, prevention, and treatment of a disease. When "The Pill" became a prescription drug in the 1950s, the definition of the word "drug" had to be changed so that federal regulations would apply to the manufacturing, distribution, advertising, and sale of the Pill. If the definition of the word "drug" had not been changed, the FDA would have had to designate pregnancy as a disease in order for federal regulations to be applicable to the Pill. Designating pregnancy as a disease would have been culturally indefensible.

Prior to *Roe v. Wade* in 1973, medical dictionaries, biology textbooks, and embryology textbooks all defined fertilization of the ovum as the onset of conception and pregnancy. The beginning of embryonic life was defined as the first cellular division of the zygote. After *Roe v. Wade* there has been a gradual alteration of the traditional, precise, and scientific definitions of conception, contraception, pregnancy, and the embryo. In our present-day culture, the beginning of conception or pregnancy is now ill-defined as the implantation of the blastocyst which occurs at about five to seven days after fertilization. This definition is not scientifically correct. Furthermore, this change in the definition of conception resulted in a change in the meaning of what constitutes a contraceptive drug. Thus the abortion of an un-implanted embryo is equated with the prevention of fertilization. The laity, unschooled in scientific terminology, now view medical abortion of un-implanted embryos as a form of contraception. A more recent re-definition of pregnancy attempts to re-define the beginning of pregnancy as the time when Beta-hCG (Beta Human Chorionic Gonadotrophin) can first be detected by the standard pregnancy tests, i.e., six to eight days after fertilization.

A new unscientific term, "pre-embryo," was invented to obscure the in vitro destruction of embryos up to fourteen days after in vitro fertilization. With this new term our culture appears to be morally accepting the harvesting of stem cells from "pre-embryos." In this form of stem cell harvesting procedure, in vitro or frozen embryos are killed in the process of obtaining stem cells for medical use in a variety of treatment protocols. Thus secular ideology promotes ambiguities of scientific terms to further a utilitarian culture of death and utilizes a ceaseless propaganda program that exploits retailored connotations of scientific terms.

Emergency Contraceptives in the Treatment of Rape Victims

The altered definition of "contraceptive" has resulted in confusion in the emergency rooms of many Catholic hospitals. This confusion revolves around whether it is licit to use emergency contraceptives in the treatment of rape victims. The confusion arises from the question "Do emergency contraceptives exert a contraceptive drug action, abortifacient drug action, or a placebo drug action when used in the treatment of rape victims?" To clear up this confusion, it is necessary to examine three areas of information: 1) the modern biology of conception; 2) the multiple drug actions of oral contraceptives; and 3) analysis of medical rape protocols and the *Ethical and Religious Directives for Catholic Health Care Services*.

The pharmacological action of a drug is classified as a contraceptive, as an abortifacient, or as a placebo depending upon the answers to three questions:

1) When does a woman become pregnant?

2) When does conception take place?

3) When does human life begin?

The most infamous answer to these questions was given by the United States Supreme Court in *Roe v. Wade*: "Because there is no agreement in the answer to these questions, we don't know."

In today's culture, a variety of politically and culturally correct answers have been invoked to justify the use of contraceptives and abortifacients. The following table provides a synopsis of the different answers given by molecular biologists, embryologists, bioethicists, theologians, clergy, physicians, women, journalists, jurists, and various secular organizations. Selective choice of an answer from this list allows a group to justify their subjective morality concerning abortion.

When Does Human Life Begin?	Time after Fertilization
We do not know, the basis of the U.S. Supreme Court's *Roe v. Wade* decision	
At fertilization	Milliseconds after sperm and ovum unite
At syngamy, fusion of the male and female pronuclei	12 to18 hours
At the zygote's first cell division	24 hours
At implantation of the embryo into the endometrium	5 to 7 days
At a detectable level of Beta-hCG	6 to 8 days
At the beginning of a heartbeat	24 days
At the beginning of brain waves	31 days
At the end of embryogenesis	56 days
At viability	154 days
At the complete birth of a baby	266 days
After neonatal health is certified as normal	48 hours after birth

The Catholic Church teaches that life, conception, and pregnancy begin at fertilization, and there is scientific evidence to support this position. At the injection of the sperm's pronucleus into the ovum, two simultaneous events take place. Within milliseconds of fertilization, a permanent electrical charge is set up across the entire external surface of the fertilized ovum's cell membrane so that no other sperm can gain entrance into a fertilized ovum. Secondly, the fertilized ovum, now called a zygote, contains the unique DNA that determines a new, unique human being. The Catholic Church's position that life, conception, and pregnancy begin at fertilization profoundly influence the moral analysis of two specific problems: 1) medical protocols used in Catholic hospital emergency departments to treat rape victims; and 2) the emerging widespread use by women of the so-called emergency contraceptives.

The position that human life begins at "complete birth" is not justified by science but by judicial mandate, which epitomizes the current transformation of our culture into the culture of death that advocates partial-birth abortion.

The Modern Biology of Conception

Knowledge of the modern day biology of conception is indispensable to analyzing the morality of different medical treatment protocols for rape victims. With rape, coitus results in several hundred million unjust aggressors deposited in the victim's vagina. The major histological structures of sperm are the flagellum, the mitochondria, the acrosome, and the haploid pro-nucleus. The whip-like actions of the flagellum allow the sperm to swim and migrate through the female uterus into the fallopian tube. The mitochondria generate the necessary biochemical energy in the form of adenosine triphosphate (ATP) for the whipping motion of the flagellum to propel the sperm. The acrosome provides the biochemical means for recognition of a species' ovum and the biochemical machinery to penetrate the ovum's cell membrane by the sperm's pronucleus.

The normal acid secretion of the vagina is the first barrier to fertilization. The acid character of vaginal secretions destroys millions of sperm. However within a few minutes of rape, a few thousand sperm escape destruction by migrating into the cervical crypts, which provide an alkaline environment. In the case of rape washing the sperm out of the vagina with saline or a spermicide does not remove nor destroy the sperm tucked away in the cervical crypts. Thus removing sperm from the vagina is an ineffective contraception procedure since it can not guarantee the prevention of fertilization. However, removal of the sperm is psychologically beneficial to the victim and is essential for obtaining evidence to convict the rapist.

Viable and mobile sperm can be stored in the cervical crypts for up to five days. While being stored in the cervical crypts, the sperm meet their second barrier to fertilization, thick cervical mucus that functions as a biological valve. Thick cervical mucus is a protective mechanism that prevents vaginal bacteria from gaining entrance into the peritoneal cavity. However, the rapid rise in estrogen level that occurs just prior to ovulation causes a hormone-mediated change in the quality of secreted cervical mucus. Thus prior to ovulation and at the peak of the estrogen level, the cervical mucus becomes watery and less viscous and forms a strong fern pattern when air dried on a microscope slide. Strongly ferning cervical mucus is easily penetrated by sperm and allow the sperm to swim through the cervix into the uterine cavity and then into the fallopian tubes. Secretions from the glands lining the uterus and the fallopian tube initiate capacitation of the sperm. Capacitation mediated by the acrosome is vital to enable a sperm to fertilize an ovum. The shortest transit time from the cervix to the ampulla of the fallopian tube has been experimentally determined to be five minutes. Thus if a rape victim has ovulated on the same day that the rape had occurred, fertilization of the ovum could result within five to ten minutes of the rape.

The size of the human ovum is just barely visible to the unaided human eye, about one half the diameter of the period at the end of this sentence. The sperm is microscopic in size and is dwarfed by the size of the ovum. To reach the cell membrane of the ovum, the sperm must physically force its way through a third barrier to fertilization, the granulosa cells that surround the zona pellucidum of the ovum. After passing through this cellular barrier, the enzymes in a sperm's acrosome digest a channel through the proteinious coating of the zona pellucidum that surround the ovum's cell membrane. Thus only capacitated sperm are able to pass through this fourth barrier to fertilization and reach the external cell membrane of the ovum. Within milliseconds of a sperm's injection of its pro-nucleus into the ovum, a permanent electrical charge spreads across the cell surface of the ovum to repel all other sperm on its surface. Then enzymes in the subcortical granules of the fertilized ovum are released into the zona pellucidum. These enzymes alter the physical nature of the zona pellucidum and render the zona pellucidum impermeable to all other sperm. The injection of the sperm's pro-nucleus is the precise moment of conception, i.e., the beginning of a new human life. Within the cell membrane of the fertilized ovum, there are now twenty-three paternal chromosomes and twenty-three maternal chromosomes.

This single, fertilized cell is now called a zygote. All future cellular events of the zygote are different stages of development of a new human life from fertilization until natural death. Within twenty-four hours the two pro-nuclei combine to form a single cell nucleus containing all forty-six chromosomes. This biological process is called syngamy. The re-combination of forty-six chromosomes form the highly specific DNA finger print that is unique to that new individual and all future cells of that individual will have that same unique DNA fingerprint. After the first cellular cleavage of the zygote, the resulting two cells are now termed an embryo. An implanted embryo will become a fetus only when organogenesis has been completed. This occurs at the end of fifty-sixth day after fertilization. The distinguishing feature of a fetus is that it has all the organs of an adult human being.

After the first cellular cleavage, the zygote becomes a two-cell embryo. This occurs in the fallopian tube where fertilization has taken place. The embryo has no biomechanical mechanism of its own to move through the fallopian tube down into the uterus. It takes another four days for the embryo to be moved into the uterine cavity though the combined action of the fallopian tube's muscular peristaltic contractions and the beating of cilia located on its internal lining. The embryo, as it enters the uterine cavity, has increased to about thirty-two cells and is named a morula. At this point in the menstrual cycle, the endometrium, the lining of the uterus, is still not in the proper state of development to allow the embryo to implant. During the next few days, the embryo continues to increase in cell number and forms a blastocyst. Development of the endometrium and the embryonic blastocyst are precisely synchronized to be favorable for implantation on about the fifth to seventh day after fertilization. The blastocyst implants itself into a suitably prepared endometrium and continues to secrete prodigious amounts of the hormone human chorionic gonadotropin (hCG). The embryo's hCG takes over the hormonal function of the maternal luteinizing hormone (LH) to ensure continuous secretion of progesterone by the corpus luteum. The hormone known as hCG is the hormone that is the basis of all pregnancy tests.

Six to eight days after fertilization, hCG reaches detectable levels in the blood and in the early morning urine sample. The six to eight day period from fertilization to a positive pregnancy test can be labeled as "a silent pregnancy." Silent pregnancy is one of the key points in the discussion whether emergency contraceptives function as contraceptives or abortifacients in the treatment of rape victims. As we will see later, the "silent pregnancy" is used by secular scientists and ethicists to obscure the abortifacient actions of the so called "emergency contraceptives."

Multiple Drug Actions of Oral Contraceptives

Oral contraceptives are female hormones that have been chemically modified by the addition of an ethinyl group to prolong their hormonal actions. Present day oral contraceptives contain low, daily doses of ethinyl estradiol (50 micrograms) and an ethinyl progestin (1 mg). Each of these two chemically altered hormones has both a contraceptive action and an abortifacient action.

Ethinyl estradiol mimics the negative feedback action of estrogen in the hypothalamus of the brain, decreasing the release of gonadotropin releasing hormone (GnRH). In the absence of GnRH, the pituitary fails to release adequate amounts of follicle stimulating hormone (FSH) and luteinizing hormonoe (LH). At low plasma levels of FSH and LH, ovarian follicles fail to develop, and ovulation does not occur. Technically this is the contraceptive action of ethinyl estradiol since fertilization cannot occur in the absence of ovulation of a ripe ovum.

Over the past fifty years the content of these chemically modified hormones in the Pill have been progressively decreased to minimize the incidence of the side effects. However, at the current low doses of ethinyl estradiol found in the present preparations of the Pill, break-through ovulation occurs more frequently when compared to the original, high dose preparations of the 1960s. Thus at these current lower doses, there is a greater likelihood of a decrease in the contraceptive drug action and an increase in the abortifacient drug action.

Ethinyl estradiol has a second pharmacological action that is abortifacient. Ethinyl estradiol alters the rate of peristaltic and ciliated propulsion of the embryo down the fallopian tube. This altered rate of propulsion results in the embryo reaching the uterine cavity at an improper point in time and thus the embryo is unable to implant into the endometrial lining.

237

The contraceptive action of the Pill's ethinyl progestin occurs due to the pharmacologically induced failure of the conversion of the thick cervical mucus to the thin cervical mucus. Only the thin, watery form of cervical mucus can be penetrated by sperm. The abortifacient action of the Pill's ethinyl progestin is the result of this hormone altering the timely development of the endometrial lining so that either implantation of the embryo is prevented or an implanted embryo is detached from the endometrial lining.

Both ethinyl estradiol and ethinyl progestin can prevent the implantation of a living embryo and then result in a "silent abortion." Women who routinely take oral contraceptives are totally unaware of the fact that the life of a developing embryo was pharmacologically terminated during this period of a silent pregnancy.

Protocol for the Treatment of Rape Victims

The following list outlines the medical treatment protocol used in hospital emergency rooms for the treatment of rape victims.

1) Treatment of injuries due to physical trauma

2) Treatment of psychological trauma

    a. Acute psychological counseling

    b. Chronic psychological counseling

3) Determination if the woman was pregnant prior to being raped

4) Prevention of sexually transmitted diseases

5) Collection of evidence for criminal prosecution (rape kit)

6) Determination if the woman is in the pre-ovulatory or post-ovulatory stage

7) Prevention of pregnancy due to the rape

In secular hospitals, step six is omitted and step seven is instituted. In secular hospitals, there is no distinction between prevention of fertilization via the contraceptive action of drugs and the termination of an embryo's life via the abortifacient action of drugs. The concept of a silent pregnancy is not recognized by secular hospitals. This failure to recognize silent pregnancies has the net result that abortion of an un-implanted embryo is synonymous with the secular definition of contraception.

In secular hospitals rape victims are treated with recent drug formulations known as emergency contraceptives. Emergency contraceptives are sold under the proprietary names of Preven, Ovral, and Plan B. These emergency contraceptives are referred to in the lay press as the "Morning After Pill" and are advertised to "prevent pregnancy" within in seventy-two hours of unprotected consensual sex or sexual assault. What is not advertised in the lay press is the fine print of the package insert for emergency contraceptives which states that these medications have a pregnancy prevention rate of only about seventy-five percent.

Preven and Ovral contain ethinyl estradiol and an ethinyl progestin, and Plan B contains only an ethinyl progestin. With these three preparations, the recommended dose is two pills taken as soon as possible after coitus and repeated twelve hours later. The chemically altered hormones in emergency contraceptives are identical to the active drugs found in the Pill, i.e., the ordinary oral contraceptive medication. Thus, the woman, within a twelve hour period, takes over four times the recommended daily dose of the oral contraceptive, and this is accompanied by severe nausea and vomiting.

The use of hormone drug preparations to prevent ovulation in rape victims as a means to avert unjust fertilization is morally licit. This may be accomplished by three hypothetical anovulatory treatment protocols. First, an intravenous infusion of gonadotropin releasing hormone antagonist at a dose which would prevent ovulation from occurring while there are still viable sperm in the cervical crypts. A second medical way of preventing ovulation may involve the oral administration of a GnRH antagonist to block the binding of gonadotropin releasing hormone to its natural pituitary receptor. Third, the use of emergency contraceptives could be used to exclusively prevent ovulation only if it can be demonstrated that ovulation has not occurred prior to the administration of an emergency contraceptive. It remains to be clinically proven which of these three anovulatory treatments is the most effective in preventing pregnancy in rape victims. Furthermore, these three different treatment modalities will have to be scrutinized to determine if there are any hidden medical or ethical prohibitions to their use in rape victims.

Assuming that a rape victim is evaluated in the emergency room within twelve hours of the rape attack, the following factors must be taken into account. The precise determination of the timing of each of the following three events, 1) the rape attack; 2) the potential fertilization phase; and 3) the administration of the first dose of the emergency contraceptive, are crucial to assigning whether the emergency contraceptive works as a contraceptive, as an abortifactient, or as a placebo (see diagram below). A serum progesterone level below 1.5 ng/ml along with a negative urine test for LH is a major indication that the woman has not yet ovulated and that the drug action of emergency contraceptives would be exclusively to prevent ovulation. Emergency contraceptives are likely to act as abortifacients if a strong ferning pattern of cervical mucus is present, if the serum progesterone level is between 1.5 and 5.9 ng/ml, if the urine test for LH is positive, and if the expected next menstrual period is more than seven days.

If the rape occurred after the potential fertilization phase, the possibility of pregnancy becomes zero even in the absence of administration of emergency contraceptives. The administration of emergency contraceptives then becomes a form of psychological treatment of the rape victim. There are two scenarios in the post-ovulatory period in which the drug action of an emergency contraceptive would be classified as a placebo. In each of these two phases of the post-ovulatory period fertilization cannot occur since the ovum, in twelve to twenty-four hours after ovulation, self-destructs. The first scenario involves a negative urine test for LH, non-ferning cervical mucus, and a serum progesterone level greater than 6 ng/ml coupled to a history of an expected menstrual period in more than seven days. The second scenario involves a negative urine test for LH, non-ferning cervical mucus, and a serum progesterone level less than 6 ng/ml coupled to a history of an expected menstrual period in less than seven days (see diagram below).

Further evidence of the ovulatory status in a rape victim can be obtained with a pelvic ultrasound examination by looking for a post-ovulatory corpus luteum or a pre-ovulatory Graafian follicle on the surface of either ovary.

Informed Consent

It is imperative that the patient, prior to being treated for rape in a Catholic hospital, gives properly informed and legally required consent. The patient must be informed that if it is determined that she has not ovulated, medication will be given to her to prevent ovulation so that fertilization and pregnancy will not occur. If it is determined that the rape occurred during the victim's ovulatory phase, abortifacient medications, including emergency contraceptives, will not be given to her and information on the location of other facilities that treat rape victims may be provided to her. If it is determined that ovulation

Pharmacological Effects of Postcoital Anovulatory Hormonal Agents

Based on a 28 Day Menstrual Cycle

	1	2	3	4	5	6	7	8	9	10	11	12	13 Ovulation	14	15	16	17	18	19	20	21	22	23	24	25	26	27	28
	<·······Menses·······>					<··········Follicular Phase··········>									<··········Luteal Phase··········>												
Days Pre & Post Ovulation						-7	-6	-5	-4	-3	-2	-1	0	+1	+2	+3	+4	+5	+6	+7							
Probability of Pregnancy[1]						<--Nearly Zero---		10%	16%	14%	27%	31%	33%	<--Nearly Zero---			<----------- Zero ----------->										
						Pre-Ovulatory							LH Surge Ovulation		Early Post-Ovulatory							Late Post-Ovulatory						
													Strong		<---Expected menses in >7 days --->							<-----Expected menses in <7 days ----->						
Cervical Mucus Property[2]						<---------- Thick ---------->							Ferning		<---------- Thick ---------->													
Urine Luteinizing Hormone[3]						-	-	-	-	-	-	+	+	+	-	-	-	-	-	-	-	-	-	-	-	-	-	-
Serum Progesterone Level[3]						<---------- < 1.5 ng/ml ---------->							>=1.5 to <=5.9 ng/ml		<---------- >=6 ng/ml ---------->							<---------- < 6 ng/ml ---------->						
Pharmacological Effects	Anovulatory Effects												Potential Abortifacient Effects							Psychological Benefit								

[1] See Allen J. Wilcox, Clarice R. Weinberg, and Donna D. Baird. "Timing of Sexual Intercourse in Relation to Ovulation—Effects on the Probability of Conception, Survival of the Pregnancy, and Sex of the Baby," *New England Journal of Medicine* 333.23 (December 7, 1995): 1517–1521.

[2] See T.W. Hilgers, G.E. Abraham, and D. Cavanagh. "Natural Family Planning:The Peak Symptom and Estimated Time of Ovulation," *Obstetrics & Gynecology* 52.5 (November 1978): 575–582.

[3] See "Interim Protocol for Sexual Assault: Contraceptive Treatment Component" from Saint Francis Medical Center in Peoria, Illinois (also known as the "Peoria Protocol").

The exact pharmacological effect of a post-coital anovulatory hormonal agent is determined by the answers to three questions:
1. Where in the menstrual cycle had the rape occurred?
2. Where in the menstrual cycle has or will ovulation occur?
3. How long after the rape is the post-coital hormonal agent given?

has occurred more than seventy-two hours prior to being raped, pregnancy due to the rape is not a possibility and anovulatory drugs are not necessary but may be administered to provide psychological comfort. Informed consent and the treatment options for these aspects of the rape protocol in Catholic hospitals are consistent with the *Ethical and Religious Directives* promulgated by the National Conference of Catholic Bishops.

A MORAL ANALYSIS OF PREGNANCY PREVENTION AFTER SEXUAL ASSAULT

PETER J. CATALDO

Among the many ethical issues associated with treatment of the victim of sexual assault that could be examined, I will focus on the narrow but critical issue of providing hormonal pregnancy prophylaxis to the victim who presents in a Catholic health care facility. This is an issue about which medical professionals, theologians, and ethicists in the Catholic moral tradition are in disagreement. Moreover, protocols for treatment of sexual assault victims among Catholic health care facilities are inconsistent with

---

This Workshop paper is adapted with minor changes from a chapter written by Albert S. Moraczewski, O.P., and me which appears in Peter J. Cataldo and Albert S. Moraczewski, O.P., "Pregnancy Prevention after Sexual Assault: III. A Moral Analysis of Pregnancy Prevention after Sexual Assault," in *Catholic Health Care Ethics: A Manual for Ethics Committees*, eds. Peter J. Cataldo and Albert S. Moraczewski, O.P. (Boston: The National Catholic Bioethics Center, 2001), 11/1–11/23 (2002 update).

each other.[1] This difference among opinion and protocols does not represent a disagreement over the moral right of the sexual assault victim to prevent a conception resulting from an assault or a disagreement about the moral reasons that can justify such an action. Rather, the difference that exists among opinions and clinical protocols revolves around two questions: 1) whether and under what circumstances hormonal intervention may have an abortifacient effect; and 2) how the possibility of an abortifacient effect should influence the moral evaluation of such an intervention. This moral analysis will presume the information from Dr. Ralph Miech in the preceding chapter[2] regarding the first question and will examine two opposing positions on the second question. The analysis will focus on an argument in favor of hormonal intervention and the moral criteria by which Catholic health care protocols could be developed.

The Role of Freedom for the Moral Analysis

A woman who is sexually assaulted has the right to defend herself against this unjust aggression before, during, and after the assault. Any semen that might have been deposited in the reproductive tract of the victim by the attacker is one of the lingering effects of the assault and can be considered part of the aggression. The woman has the right to defend herself against this effect and the possibility that it will lead to fertilization. It has been long recognized in the Catholic moral tradition that if it is morally justifiable for a woman to take measures to prevent a sexual attack, then it is justifiable for her to prevent any continuation of the same attack.[3] Every aspect of

---

[1]See Steven S. Smugar, M.D., Bernadette J. Spina, and Jon F. Merz, J.D., "Informed Consent for Emergency Contraception: Variability in Hospital Care of Rape Victims," *American Journal of Public Health* 90.9 (September 2000): 1372–1376.

[2]"Rape Protocols and Abortifacient Actions of Oral Contraceptives," 231–242.

[3]See William E. May, *Catholic Bioethics and the Gift of Human Life* (Huntington, IN: Our Sunday Visitor Publishing Division, Our Sunday Visitor, Inc., 2000); Thomas J. O'Donnell, S.J., *Medicine and Christian Morality*, 3d rev. ed. (New York: Alba House, 1996); Germain Grisez, *Living a Christian Life*, vol. 2, *The Way of the Lord Jesus* (Quincy, IL: Franciscan Press, 1993); Edward J. Bayer, *Rape within Marriage: A Moral Analysis Delayed* (Lanham, MD: University Press of America, 1985); John T. Noonan, Jr., *Contraception: A History of Its Treatment by the Catholic Theologians and Canonists* (New York: New American Library, 1965).

the act (including the attacker's semen and the risk of fertilization) is forced upon her, and is against her free choice of the will and free consent. As an act of self-defense, the woman may take measures to prevent fertilization. She may stop the lingering effect of the attack by not allowing her fertility to be integrated with that of her attacker.[4]

A question arises. How can the Catholic moral tradition consistently hold that the deliberate suppression of fertility is morally unacceptable in some cases but not in others? The morally decisive difference between deliberately suppressing fertility as a defensive measure against unjust aggression and deliberately suppressing fertility only for the sake of preventing fertilization (e.g., direct contraception) rests on these facts: 1) the suppression of fertility in the case of sexual assault is in response to an involuntary sexual act; and 2) the suppression of fertility in the case of contraception is part of a voluntary sexual act. Thus, the presence or absence of voluntariness (i.e., free consent to the act of intercourse) is an essential condition that determines a moral difference between what are otherwise two similar physical acts, namely, taking hormonal medication to suppress ovulation in the context of freely chosen intercourse on the one hand, and on the other, taking the same medication after a sexual assault.

The Church's prohibition against contraception and direct sterilization is, among other things, premised on the fact that those for whom the prohibition applies freely engage in sexual intercourse.[5] If an act of sexual intercourse is voluntary, then neither contraceptives nor surgical sterilization may be used. If one is free to choose regarding the use of the reproductive powers, then one is under the moral obligation to act in accordance with the inherent good of those powers and avoid what is evil in fulfillment of that freedom. The voluntary use of the reproductive powers must therefore be reserved for genuine con-

[4]See Grisez, *Living a Christian Life*, 512; Bayer, *Rape Within Marriage*, 9.

[5]See Benedict M. Ashley, O.P. and Kevin D. O'Rourke, O.P., *Health Care Ethics: A Theological Analysis* (Washington, D.C.: Georgetown University Press, 1997), 303; Grisez, *Living a Christian Life*, 512; Francis J. Connell, C.SS.R., "Is Contraception Intrinsically Wrong?" *American Ecclesiastical Review* 150.6 (June 1964): 434–439; idem, "The Sterilization of a Retarded Girl," *American Ecclesiastical Review* 154.4 (April 1966): 280–281; L.L. McReavy, "The Dutch Hierarchy on Marriage Problems," *Clergy Review* 49 (February 1964): 113–115.

jugal acts in marriage. Pope Pius XII explicitly stated the link between the voluntary use of the reproductive powers and their inherent goal: "[T]he Creator Himself, for the good of the human race, has indissolubly bound up the voluntary use of those natural energies [of sexuality] with their intrinsic purpose."[6] It follows from this statement that the Church's prohibition against contraception also rests on the same foundation of freedom of action.

This point is made in the following texts from the Congregation for the Doctrine of the Faith (CDF):

> [S]terility intended in itself is not oriented to the integral good of the person as rightly pursued, "the proper order of goods being preserved," inasmuch as it damages the ethical good of the person, which is the highest good, since it deliberately deprives *foreseen and freely chosen sexual activity* of an essential element. (Emphasis added.)[7]

In its *Declaration on Certain Questions concerning Sexual Ethics* the CDF again indicates the role of freedom in the Church's prohibition against use of the reproductive powers outside conjugal relations in marriage:

> [T]he deliberate use of the sexual faculty outside normal conjugal relations essentially contradicts the finality of the faculty. For it lacks the sexual relationship called for by the moral order, namely, the relationship which realizes "the full sense of mutual self-giving and human procreation in the context of true love." *All deliberate exercise of sexuality must be reserved to this regular relationship.*[8] (Emphasis added.)

Free use of the reproductive powers carries the obligation to use them only for their inherent ends, namely, their procre-

---

[6]Pope Pius XII, "Christian Principles and the Medical Profession," allocution to the Italian Medical-Biological Union of St. Luke (November 12, 1944) in *The Human Body*, ed. Monks of Solesmes (Boston: Saint Paul Editions, 1960), 51–65, at 62.

[7]Congregation for the Doctrine of the Faith, "Reply of the Sacred Congregation for the Doctrine of the Faith on Sterilization in Catholic Hospitals," March 13, 1975, in *Commentary on the Reply of the Sacred Congregation for the Doctrine of the Faith on Sterilization in Catholic Hospitals*, National Conference of Catholic Bishops (Washington, D.C.: National Council of Catholic Bishops, 1983), n. 1, quoting Pope Paul VI, *Humanae vitae*, n. 10.

[8]Congregation for the Doctrine of the Faith, *Declaration on Certain Questions Concerning Sexual Ethics*, December 29, 1975 (Washington, D.C.: United States Catholic Conference, 1976), n. 9, quoting Vatican Council II, *Gaudium et spes*, n. 51.

ative and unitive meanings. If a sexual act is forced upon a woman and is not freely engaged in, then the prohibition against deliberately suppressing the procreative dimension of the act does not apply. From a moral point of view the act of preventing fertilization in this situation is different in kind from an act of contraception. The absence of the essential condition of freedom in a sexual assault makes the prevention of fertilization after the assault an act of self-defense rather than a contraceptive act (in the moral sense). Given that it is morally justifiable to suppress ovulation as an act of self-defense against sexual assault, is it morally acceptable to achieve this by hormonal intervention?

The Moral Status of Postcoital Anovulatory Hormonal Drugs

The disagreement already mentioned among physicians, theologians, and ethicists who are in accord with the magisterium of the Church is specifically over whether the use of certain drugs (such as Ovral and Preven), during identifiable infertile phases of the victim's ovulation cycle, constitutes a morally legitimate act of self-defense, or whether their use during these phases constitutes the moral equivalent of abortion by preventing implantation. Thus, there is disagreement as to whether compliance with all parts of Directive 36 of the *Ethical and Religious Directives for Catholic Health Care Services*[9] is morally possible at the present time. The directive states:

> Compassionate and understanding care should be given to a person who is the victim of sexual assault. Health care providers should cooperate with law enforcement officials, offer the person psychological and spiritual support and accurate medical information. A female who has been raped should be able to defend herself against a potential conception from the sexual assault. If, after appropriate testing, there is no evidence that conception has occurred already, she may be treated with medications that would prevent ovulation, sperm capacitation, or fertilization. It is not permissible, however, to initiate or to recommend treatments that have as their purpose or direct effect the removal, destruction, or interference with the implantation of a fertilized ovum.

---

[9]United States Conference of Catholic Bishops, *Ethical and Religious Directives for Catholic Health Care Services*, 4th ed. (Washington, D.C.: United States Conference of Catholic Bishops, 2001).

The Directive allows for medications that would "prevent ovulation." This point, together with the proscription against preventing implantation in the last sentence of the Directive, imply that the health care provider must have a prudential or moral certitude (based upon current medical and pharmacological knowledge) that the medication will more likely prevent ovulation rather than primarily act as an abortifacient for any case in which such a drug is administered. Therefore, reliable testing to identify the ovulatory phase of the sexual assault victim is critical for the proper implementation of Directive 36.[10] It is at this juncture that the moral opinions about the use of medications to prevent ovulation in the treatment of sexual assault victims diverge.

An Argument against the Use of Postcoital Hormonal Drugs

The argument against the use of postcoital hormonal drugs for their anovulatory effect may be summarized in this way:

The use of postcoital hormonal drugs administered only for their anovulatory effect is not morally permissible, because

A. The evidence indicates that these drugs can also prevent implantation of the embryo, which is the moral equivalent of abortion, and

B. The principle of the double effect does not validly apply, if attempted.

It can be argued that the inability to know whether hormonal drugs will only prevent ovulation (and not the implantation of an embryo) makes their use morally impermissible. As long as there is a possibility that the drug can have the post-fertilization effect of preventing implantation, the drug can cause the death of a child conceived which would be morally unacceptable. If it is not known that a child has been conceived in any given case, even the risk of causing the death of a child is

---

[10]The St. Francis Medical Center Interim Protocol for Sexual Assault (St. Francis Medical Center, Peoria, Illinois), analyzed by Dr. Miech in the preceding chapter and mentioned below, is one way of obtaining the requisite moral certitude. The Seton Health Care Network Emergency Departments Administrative Standard (Austin, Texas) is another protocol which provides for sufficient moral certitude (hereafter, "Seton protocol"). The fact that these points regarding the need for reliable testing are implicit in the Directive and not explicit indicates that a clarification of the text is needed and would be helpful.

not morally justified, especially in light of the fact that the life of the woman is not at stake. To employ such hormonal treatment with the knowledge (in advance) that there is a realistic possibility that the drug could have an abortifacient effect is to be morally responsible for causing the death of the child should this effect occur. Moreover, to risk the realistic possibility of an abortifacient action is itself a morally reprehensible act even if no child is killed.

Therefore, because the possible abortifacient effect of these drugs is known and accepted by the majority of the medical community, Catholic health care institutions may not use these drugs in the treatment of sexual assault victims. Moreover, one cannot appeal to the principle of the double effect because it does not validly apply in the case. There are not two concurrent effects (one good and the other bad) caused by the act of administering the drug. Rather, in each particular case, only one of two possible effects occurs—either suppression of ovulation or prevention of implantation.[11]

An Argument in Favor of the Use of Postcoital Hormonal Drugs

Our argument in favor of using postcoital hormonal drugs for their anovulatory effect may be summarized in this way:

The use of postcoital hormonal drugs administered only for their anovulatory effect is morally permissible, because

A. The evidence indicates that these drugs act only as anovulants in the majority of cases and do not necessarily have an abortifacient effect in the remainder of cases, and

B. The remote possibility of a secondary mechanism occurring which might prevent implantation during an anovulatory phase is morally justifiable under the proper

---

[11]See O'Donnell, *Medicine and Christian Morality*, 196–197; Eugene F. Diamond, M.D., "Rape Protocol," *Linacre Quarterly* 60.3 (August 1993): 8–19, at 13–14, idem, "Ovral in Rape Protocols," *Ethics & Medics* 21.10 (October 1996): 1–2, at 2; and Steven P. Rohlfs, "Pregnancy Prevention and Rape: Another View," *Ethics & Medics* 18.5 (May 1993): 1–2. Arguments that the principle of the double effect does not validly apply for cases in which drugs like Ovral are used have been largely based upon the assumption that there is no reliable method of determining the ovulatory phase of the sexual assault survivor. See May, *Catholic Bioethics*, 149, note 44.

circumstances and conditions of moral certitude using a three-moral-fonts approach.

The argument we use applies the traditional three fonts or sources of the moral act to the question rather than the principle of the double effect. We believe that the question is not one that can be judged by the principle of the double effect in the first place, but for reasons other than what have been mentioned earlier in the argument against the use of postcoital hormonal drugs. Understanding why the principle does not apply will be instructive for the argument in favor of hormonal intervention.

The principle of the double effect requires the foreseeing of two effects of an act (one effect being good and the other evil) that will occur as a result of the act. This prerequisite condition for applying the principle does not obtain in the question of anovulatory hormonal treatment of a victim of sexual assault. The evil effect of preventing the implantation of a fertilized ovum as a result of using anovulatory hormonal drugs only during an infertile phase of the sexual assault victim, while possible, is improbable.[12]

---

[12]Studies have indicated that twenty to thirty percent of pregnancies are prevented (or halted) when using Ovral during the preovulatory phase, not by suppression of ovulation, but by any one of a number of possible causes including prevention of implantation. See W.Y. Ling et al., "Mode of Action of dl-Norgestrel and Ethinylestradiol Combination in Postcoital Contraception," *Fertility and Sterility* 32.3 (1979): 297; A.J. Wilcox et al., "Timing of Sexual Intercourse in Relation to Ovulation: Effects on the Probability of Conception, Survival of the Pregnancy, and Sex of the Baby," *New England Journal of Medicine* 333.23 (December 7, 1995): 1518; M. Swahn et al., "Effect of Post-Coital Contraceptive Methods on the Endometrium and the Menstrual Cycle," *Acta Obstetricia et Gynecologica Scandinavica* 75 (1996): 738. See also Walter L. Larimore, M.D., and Joseph B. Stanford, M.D., M.S.P.H., "Postfertilization Effects of Oral Contraceptives and Their Relationship to Informed Consent," *Archives of Family Medicine* 9.2 (February 2000): 126–133. It is generally reported that the rate of pregnancy after rape is zero to four percent. See R.B. Everett and G.F. Jimerson, "Rape Victim: A Review of 117 Consecutive Cases," *Obstetrics and Gynecology* 50.1 (July 1977): 88; S.K. Makhorn, *Pregnancy and Sexual Assault: Psychological Aspects of Abortion* (Washington, D.C.: University Publications, 1979). If the 20 to 30% rate is multiplied by the rate of pregnancy after rape, the chance of an abortifacient effect in a sexual assault survivor should be 1.2% or less (even less under the restrictions of the Catholic protocols for pregnancy prevention after sexual assault—mentioned above in note 10 and below).

If one of two possible effects is improbable, then that effect does not bear the sort of causal relation to the act that would warrant an evaluation through an application of the principle.[13] But the fact that a particular effect is highly probable indicates a stronger causal connection to the particular act, which in turn has a corresponding influence on the moral status of the act and makes the effect morally significant. The unlikely probability that an abortifacient effect will occur as a result of receiving an anovulatory hormonal medication during an infertile phase is, because of its improbability, disqualified as an effect that would legitimately trigger an application of the principle of the double effect. There is no morally significant evil effect which could be factored into a double effect analysis because the evidence indicates that, within certain circumstances, prevention of implantation is improbable, and because evidence to the contrary is inconclusive. Moreover, if such an evil effect is improbable, that effect cannot be foreseen in any morally meaningful sense of foreseeing required by the principle. If there are not sufficient grounds confidently to foresee an evil effect, then the one condition needed to invoke the principle of the double effect does not obtain. Therefore, a different sort of moral evaluation is needed, namely, one that applies the traditional three fonts or sources of a moral act: the object, intention, and circumstances of an act.

The sexual assault victim commits an act containing both physical and moral components which occur during the treatment of the victim.[14] Physically and biologically there is an external act of receiving an anovulatory drug, which morally is an act of self-defense. This external act, as an act of self-defense, is a morally good kind of act. The *moral object* of the act may be identified as self-defense against the consequences of the unjust sexual aggression of the attacker. However, the fact that the moral object of the act is good does not by itself justify the acts. The intention and circumstances of the act must also be morally right.

The intention of the victim must be to suppress ovulation in order to prevent an unjust pregnancy and not to cause the

---

[13]This weak causal relation between act and effect is essentially why a double effect analysis is not required for common, ordinary actions like the proper operation of a motor vehicle, as a result of which operation deaths nonetheless occur at a definite rate.

[14]A similar analysis can be made for the parallel acts of the health care professional.

death of a child conceived, if fertilization has occurred. The good moral object notwithstanding, *if* the intention of the victim also includes intending the possible abortifacient effect of the drug, then the act is immoral, because the victim would be intending the killing of an innocent human being. This sort of intention must be distinguished from foreseeing the possibility that the drug might prevent implantation. The fact that an improbable abortifacient effect may be foreseen is not part of the intention. Foreseeing and intending are acts of two related but different human powers. To foresee the possibility of something is an act of the intellect and to intend something is an act of the will. Thus, to foresee (know) that an abortifacient effect might occur is not necessarily to will (intend) it.

The moral object of the act might be good and the intention of the victim might be right but if the circumstances surrounding the act are not in due proportion, i.e., if they are morally defective, then the act is immoral. Traditionally, the circumstances of a human act comprising this third moral font include who, what, by what means, where, why (the intention just discussed), how, and when. Here is an example illustrating all seven circumstances and their traditional terms as they pertain to a typical case of a nurse administering pain medication:

1. *Quis* (who): a registered nurse;
2. *Circa quid* (about what) or simply *quid* (what): a) "by reason of quantity," therapeutic dose of a medication; b) "by reason of quality," morphine; c) "by reason of the effect," direct alleviation of pain;
3. *Quibus auxiliis* (by what means): an intravenous line;
4. *Ubi* (where): health care facility;
5. *Cur* (why): to provide palliative care to a dying person;
6. *Quomodo* (how): competently and with compassion; and
7. *Quando* (when): at the time when pain is experienced or in anticipation of the pain.

These same types of circumstances may be applied to anovulatory hormonal intervention for the treatment of a victim of sexual assault:

1. Who: the victim and the attending health care professionals;
2. What: receiving the drug, e.g., Ovral and its actions, namely, suppression of ovulation, alteration of cervical

mucus, and changes to the endometrium and fallopian tubes;

3. By what means: oral ingestion of the medication;

4. Where: emergency department of a Catholic hospital;

5. Why: to prevent fertilization from an unjust act;

6. How: with a pregnancy test; use of tests for luteinizing hormone and serum progesterone levels to determine the ovulatory phase of the victim; victim's ovulation history; delivered in acute doses; with a moral certitude about the ovulatory phase of the victim; and

7. When: only during the infertile phases of the victim's ovulation cycle.

In order for the anovulatory hormonal intervention to be justified, these circumstances must be in due proportion to the possibility that the drug might prevent implantation. This means that the circumstances of the act must be such that they produce a moral certitude that the hormonal intervention will not prevent the implantation of a fertilized ovum. The relevant circumstances that coalesce to provide this certitude are: a) the primary mechanism of the drug when it is given is the inhibition of ovulation;[15] and b) it is given in acute doses when, and only when, an oocyte is not available for fertilization, i.e., during either the preovulatory phase, past the early postovulatory phase, or in the late postovulatory phase of the woman's cycle. As is illustrated by the St. Francis Medical Center Interim Protocol for Sexual Assault (hereafter, "St. Francis Medical Center protocol") presented by Dr. Miech, identification of these phases is possible in the clinical setting. At a minimum, a pregnancy test must be performed (which if positive would prohibit use of the drug) along with a urine test to determine the luteinizing hormone level.[16] The victim's ovulation history should also be taken. Administered under these circumstances the drug acts only as an anovulant in the majority of cases and may or may not have an abortifacient effect in the remainder.

---

[15]See Larimore and Stanford, "Postfertilization Effects of Oral Contraceptives," for a good description of the various mechanisms of oral contraceptives.

[16]Since the time that this paper was originally delivered, the Seton protocol has been developed which shows that the serum progesterone test is not necessary in all cases to secure the requisite moral certitude for administering an anovulatory hormonal agent.

Thus, even if there is a remote chance that the drug can have an abortifacient effect under the circumstances described, the three fonts or sources of a good moral act are nevertheless fulfilled.[17] If the proper testing has been performed to determine the ovulatory phase of the victim and it is known that the victim is in the infertile phases identified above, then the disposition of the circumstances indicate that it is morally certain that an abortifacient effect will be avoided. This conclusion, together with a morally good object and intention, leads to the morally certain judgment that the administration of the drug is justified.

Moral Certitude in Hormonal Intervention

The Catholic moral tradition has always recognized that certitude in a judgment about human action is not the same as what is required for certitude in the physical sciences or in mathematics. The reason for the difference is that certitude is proportionate to the nature of the particular subject matter known. This principle was first articulated by Aristotle in his *Nicomachean Ethics*, was adopted by Saint Thomas Aquinas, [18] and has guided the Catholic moral tradition. Referring to the nature of ethics, Aristotle explains that "Our discussion will be adequate if it has as much clearness as the subject-matter admits of, for precision is not to sought for alike in all discussions." He states that we should "look for precision in each class of things just so far as the nature of the subject admits."[19] Because moral judgment concerns the practical realm of acting, it must proceed on the basis of things that are often unclear, changeable, and variable (though not completely so). In con-

---

[17]For other moral arguments in favor of anovulatory hormonal treatment see Orville N. Griese, *Catholic Identity in Health Care: Principles and Practice* (Braintree, MA: The Pope John Center, 1987), 336–337; Ashley and O'Rourke, *Health Care Ethics*, 303–307; Joseph J. Piccione, "Rape and the Peoria Protocol," *Ethics & Medics* 22.9 (September 1997): 1–2.

[18]For example, see Thomas Aquinas, *Commentary on Aristotle's Nicomachean Ethics*, trans. C.I. Litzinger, O.P., (Notre Dame, IN: Dumb Ox Books, 1993), Bk. 1, lect. 3, especially n. 32, 11–12: "Now the matter of moral study is of such a nature that perfect certitude is not suitable to it."

[19]See Aristotle, *Nicomachean Ethics*, trans. W.D. Ross in *The Basic Works of Aristotle*, ed. Richard McKeon (New York: Random House, 1941), I.3, 1094b12–29.

trast, a scientific conclusion is based upon things that are largely (though not completely) predicable, universal, and uniform. This difference in subject-matter has been traditionally characterized as a difference resulting in what is known as moral certitude for practical judgments and scientific or mathematical certitude for scientific judgments.

The certitude of a moral judgment about human action need not be, and indeed cannot be, absolute. A morally certain judgment is one that does not include any prudent fear of erring, but does include reasons that might militate against the truth of what is judged. However, these mitigating reasons are indecisive and are eclipsed by evidence to the contrary. This fact about moral certitude does not preclude strict certainty about assents to relevant moral principles that a person makes leading up to a particular moral judgment, e.g., assenting with strict certainty to the truth of the moral principle that innocent human life ought never be directly taken. However, assent to the judgment that this action, under these circumstances, does not directly take innocent human life and is morally acceptable may contain inconclusive evidence to the contrary. Such a judgment is made with moral certitude. The certitude of this moral judgment is not the same with which we assent, for example, to the truths of mathematics, like the Pythagorean theorem which states that for a two-dimensional surface, the square of the hypotenuse of a right triangle equals the sum of the squares of the other two sides. However, the certitude of the moral judgment in question is sufficient for the nature of the subject matter that is judged.

A judgment that is made with moral certitude has the trait of being what the Catholic moral tradition calls a "solid probability."[20] A three-fold test may be applied to determine if a moral

---

[20]By utilizing some of the categories and criteria for moral certitude from the Catholic moral tradition of casuistry, we are not thereby endorsing casuistry, suggesting that the categories of casuistry dominate Catholic bioethics, or suggesting that casuistry be returned to its former glory. Romanus Cessario, O.P., "Towards an Adequate Method for Catholic Bioethics," *National Catholic Bioethics Quarterly* 1.1 (Spring 2001): 51–62, has recently demonstrated the deficiencies of casuistic theology as it relates to Catholic bioethics. What is needed is a Catholic bioethics that properly integrates the complementarity between Divine Law and human liberty found in Pope John Paul II's *Veritatis splendor* (August 6, 1993) with some of the practical guides from the casuistic tradition. The highly complex and often uncertain subject-matter of some questions in bioethics can be aided by an

judgment is solidly probable: 1) the argument in its favor would be considered significant by a recognized authority; 2) there can be no decisive argument from authority or reason against the judgment; and 3) the arguments in favor of the judgment are not all satisfactorily refuted by the arguments for the contrary judgment.[21] The moral acceptability of receiving an anovulatory hormonal drug (such as Ovral) only during an infertile phase of the ovulation cycle, which has been confirmed by proper testing, appears to meet the criteria for solidly probable moral certitude. First, arguments like what we have proposed in favor of hormonal intervention have been considered significant by notable theologians who are in accord with the magisterium.[22] Second, there is no decisive evidence from authority or reason against the claim that, in any given case under the circumstances described, the hormonal drug is likely not to act as an abortifacient and may be licitly used in those circumstances.[23] Finally, the arguments against the moral permissibility of hormonal intervention during the anovulant phases do not pose

---

application of notions like the nature of moral certitude, or principles like the principle of cooperation in an effort to secure the good of the human person in health care, especially Catholic health care.

[21]John A. McHugh, O.P., and Charles J. Callan, O.P., *Moral Theology: A Complete Course*, vol. 1, rev. by Edward P. Farrell, O.P. (New York: Joseph F. Wagner, 1958), 259–260.

[22]For example, Ashley and O'Rourke, *Health Care Ethics*; Griese, *Catholic Identity in Health Care*.

[23]For opposing medical and ethical positions from prolife physicians and a pharmacological expert on the question of the abortifacient effect of oral contraceptives see Larimore and Stanford, "Postfertilization Effects of Oral Contraceptives"; Larimore and Randy Alcorn, M.A., "Using the Birth Control Pill Is Ethically Unacceptable" in *The Reproduction Revolution: A Christian Appraisal of Sexuality, Reproductive Technologies, and the Family*, eds. John F. Kilner, Paige C. Cunningham, and W. David Hager (Grand Rapids, MI: Wm. B. Eerdmans Publishing Co., 2000), 179–191; Susan A. Crockett, M.D., et al., "Using Hormone Contraceptives Is a Decision Involving Science, Scripture, and Conscience" in *The Reproduction Revolution*, Kilner et al., eds, 192–201; Joel E. Goodnough, M.D., "Redux: Is the Oral Contraceptive Pill an Abortifacient?" *Ethics & Medicine* 17.2 (Summer 2001): 37–51; John Wilks, B., Pharm, M.P.S., "Response to Joel Goodnough M.D., 'Redux: Is the Oral Contraceptive Pill an Abortifacient?'" *Ethics & Medicine* 17.2 (Summer 2001): 103–109; and William F. Colliton, M.D., "Response to Joel Goodnough MD, 'Redux: Is the Oral Contraceptive Pill an Abortifacient?'" *Ethics & Medicine* 17.2 (Summer 2001): 110–113.

alternative explanations which are more satisfactory. In fact, as shown above, there is significant evidence about the anovulatory effectiveness of the drug for the majority of cases, and there is uncertainty about whether prevention of implantation occurs in any of the remaining cases treated under the restrictive circumstances described above.[24]

It may be objected that the Catholic moral tradition always recognized an obligation to follow the safer course of action when there is doubt about a matter of great value which would be put at risk, such as the loss of innocent human life. According to the tradition, the safer course must be pursued rather than expose oneself or another to great harm.[25] There are two classic examples of this obligation to take the safer course: 1) the case of a person suspecting that a drink he or she wishes to have is poisonous; and 2) the case of a hunter's doubt about whether an object that he or she is set to shoot is a human being. The person in these cases is obliged to take the safer option of refraining from acting. From this perspective it may be argued that any doubt about the possibility of an abortifacient effect from anovulatory hormonal intervention is analogous to the classic cases of the rule always to follow the safer course. The risk to innocent human life from the hormonal intervention is analogous to the risk in the classical cases and, therefore, warrants the safer course of not providing the intervention.

This objection is not applicable for two reasons. First, the case of anovulatory hormonal treatment is essentially different from the classic cases. In the classic cases the fact that an object (i.e., the drink, the animal) exists is not in question. The doubt does not concern *that* there is an object, but *what* the object is. The poisonous drink case poses doubt about what type of drink is in the glass, not whether there is a substance for drinking. The hunter case poses doubt about what sort of ani-

---

[24]Moreover, authors such as Chris Kahlenborn, Joseph B. Stanford, and Walther L. Larimore ("Postfertilization Effect of Hormonal Emergency Contraception," *Annals of Pharmacotherapy* 36.3 [March 2002], 465–470), who advise the prohibition of hormonal intervention in restrictive Catholic protocols based upon the possibility of an abortifacient effect, do not describe this effect in any given case as being certain or even probable, but rather admit that the effect "may" occur and indeed describe it as a "rare possibility" (at 468).

[25]See, for example, McHugh and Callan, *Moral Theology*, n. 641, (231); n. 661, (240–241); n. 678, (247–248).

mal or object is in sight, not whether there is a target for shoot-
ing. The doubt in these cases is generated in the first place
because it is strictly certain that an object exists. It is this
particular certainty that gives rise to a doubt about what the
nature of that object or entity might be. In the case of anovulatory hormonal treatment the doubt concerns whether there
is an object, namely, a newly conceived being in the woman's
reproductive tract. If there is an object, its nature is not in
doubt—it is a human being. The question in the case of anovulatory hormonal treatment is not whether the physical object which exists is a human being, but rather whether any
human being exists at all (relative to the person's act). In the
case of the hunter, the fact that a target exists (which might
be human) is by itself an added reason for taking the safer
option. However, in the anovulatory hormonal intervention
case there is less reason to believe that a human being will
be harmed compared to the classic cases, because the doubt
in the anovulatory hormonal intervention case is about
whether there is any entity at all which might be affected by
the action. The evidence indicates that in a significant majority of cases studied (at least seventy to eighty percent) it is
likely that there is no individual human being who could be
harmed by the intervention, because the intervention effectively prevents conception in those cases. With respect to the
remaining twenty to thirty percent of cases, there is no definite or conclusive evidence that a human individual is present
(see note 12 above). Moreover, the probability that no human
individual is present would seem to increase for women treated
under the restrictive conditions of the St. Francis Medical
Center or Seton protocols.[26]

This fact about what the evidence indicates is related to
the second reason why the traditional rule to take the safer
course does not apply to the case of anovulatory hormonal treatment. The safer course rule was often applied in cases of "nega-

---

[26]The conclusion of Kahlenborn, Stanford, and Larimore that
"Catholic hospitals that do allow hormonal EC [emergency contracep-
tion] use prior to ovulation may wish to reassess their policies" in
light of the authors' claim that the use of emergency contraception
"does not always inhibit ovulation even if used in the preovulatory
phase" ("Postfertilization Effect," 468), implies that a hormonal intervention cannot be licitly administered if there is not one hundred percent certainty that no human being will be harmed. This assumption
does not take into account the Catholic tradition on moral certitude.

tive doubt."[27] A person's conscience was considered to be in negative doubt if there were no reasons of any significance on either side of two opposing possible actions. A conscience in this state would be resolved by a governing obligation if fulfilling that obligation was the safer course. However, in the case of anovulatory hormonal intervention there is not an equal lack of significant reasons for acting or not acting. Administered under the restrictive conditions found in the St. Francis Medical Center or Seton protocols, there are significant reasons to believe that another human being will not be harmed by the hormonal intervention. This fact indicates that the case of anovulatory hormonal intervention is not an instance of traditional negative doubt, which might otherwise be settled by an obligation not to act because this would be the safer course. Therefore, because of the improbability of harming an innocent human being and the absence of negative doubt, the traditional rule always to follow the safer course does not apply to the restrictive anovulatory hormonal intervention of protocols like the St. Francis Medical Center or Seton protocols and should not pose an obstacle to the intervention.

In conclusion we hold that the acts of administering and taking an anovulatory hormonal drug such as Ovral for the medical treatment of a sexual assault victim can be morally justified by a proper fulfillment of the traditional three moral fonts. If the moral object of the act is self-defense against the unjust sexual aggression of the attacker and the intention is to suppress ovulation in order to prevent an unjust pregnancy (and not to cause the death of a child which may have been conceived) and the disposition of the morally relevant circumstances (such as what drug is given and when) is consistent with the object and intention, then the administration of the drug may proceed. If all these conditions are fulfilled, the morally certain judgment can be made that the hormonal intervention is not preventing implantation and is morally justified.

This conclusion represents the considered theological and ethical opinion of Albert S. Moraczewski, O.P., and myself. If this opinion is found in error by the magisterium, or is found to be in any way inconsistent with the teaching of the magisterium, then we will gladly retract the opinion and uphold the teaching of the magisterium.

---

[27]See, for example, McHugh and Callan, *Moral Theology*, n. 661, 240–241.